Cambridge Elements ≡

Elements in the Philosophy of Science
edited by
Jacob Stegenga
University of Cambridge

MODELING SCIENTIFIC COMMUNITIES

Cailin O'Connor
University of California

CAMBRIDGE
UNIVERSITY PRESS

Shaftesbury Road, Cambridge CB2 8EA, United Kingdom

One Liberty Plaza, 20th Floor, New York, NY 10006, USA

477 Williamstown Road, Port Melbourne, VIC 3207, Australia

314–321, 3rd Floor, Plot 3, Splendor Forum, Jasola District Centre,
New Delhi – 110025, India

103 Penang Road, #05–06/07, Visioncrest Commercial, Singapore 238467

Cambridge University Press is part of Cambridge University Press & Assessment,
a department of the University of Cambridge.

We share the University's mission to contribute to society through the pursuit of
education, learning and research at the highest international levels of excellence.

www.cambridge.org
Information on this title: www.cambridge.org/9781009454087

DOI: 10.1017/9781009359535

First published 2023

A catalogue record for this publication is available from the British Library.

ISBN 978-1-009-45408-7 Hardback
ISBN 978-1-009-35954-2 Paperback
ISSN 2517-7273 (online)
ISSN 2517-7265 (print)

Modeling Scientific Communities

Elements in the Philosophy of Science

DOI: 10.1017/9781009359535
First published online: November 2023

Cailin O'Connor
University of California

Author for correspondence: Cailin O'Connor, cailinmeister@gmail.com

Abstract: This Element will overview research using models to understand scientific practice. Models are useful for reasoning about groups and processes that are complicated and distributed across time and space, that is, those that are difficult to study using empirical methods alone. Science fits this picture. For this reason, it is no surprise that researchers have turned to models over the last few decades to study various features of science. The different sections of the Element are mostly organized around different modeling approaches. The models described in this Element sometimes yield takeaways that are straightforward, and at other times more nuanced. The Element ultimately argues that while these models are epistemically useful, the best way to employ most of them to understand and improve science is in combination with empirical methods and other sorts of theorizing.

Keywords: scientific communities, philosophy of science, models, metascience, agent-based modeling

ISBNs: 9781009454087 (HB), 9781009359542 (PB), 9781009359535 (OC)
ISSNs: 2517-7273 (online), 2517-7265 (print)

Contents

1 Introduction

On a naive view, science involves a set of practices that unerringly march toward the truth. Scientists use the best methods available to gather evidence. They reason dispassionately about this evidence and change their beliefs and theories accordingly. They share their data freely and widely and listen carefully and fairly to the findings of other scientists. When they discover that their methods or theories are flawed, they abandon them for better ones.

Of course, the reality is messy and imperfect. Scientists are humans, and, like all human enterprises, science has successes and failures, good practices and poor ones. Foibles of human psychology impact science at every stage of the process, from grant seeking, to hypothesis choice, to evidence gathering, to theory generation, to argumentation and publication.

This messy reality means two things. First, to understand the workings of science, researchers must study it as a human enterprise. Second, this study has the potential to improve scientific practice. While science is imperfect, it is also often self-reflective and self-correcting. By studying science, it is possible to make discoveries about which features of the scientific process are the most successful (or the most problematic) and make changes accordingly.

Starting in the mid-twentieth century, theorists have engaged in just this sort of study under the headings of "philosophy of science," "sociology of science," "the science of science," and, more recently, "metascience." Researchers across the social sciences, philosophy, and even STEM disciplines like engineering, biology, and computer science have investigated their own practices, and the practices of their colleagues, to see how science works and how it might work better.

The goal of this Element will not be to overview this broad ranging literature but to focus on one part of it – research using models to understand scientific communities. In recent decades, the use of models to study human behavior has become increasingly popular. Especially as researchers gain access to more and more computational power, it has become clear that mathematical and computational representations of human groups have the capacity to elucidate a wide range of phenomena. In particular, models are useful for reasoning about groups and processes that are complicated and distributed across time and space, that is, those that are difficult to study using empirical methods alone. Science fits this picture. Scientific theory change, for example, often happens over significant time spans and involves thousands of interactions between hundreds of researchers performing hundreds of experiments. For this reason, it is no surprise that researchers have turned to models to study various features of science. As we will see, these models can play many roles in the study of scientific communities.

This Element is a short overview. For most of the models described, I will not go into much mathematical detail, instead focusing on general descriptions and take-aways. Notably, this Element will not give an in-depth survey of other theoretical and empirical work on science, except inasmuch as this work is relevant to the models of science discussed. It also will not survey the large literature using models to study social epistemology – the spread and development of ideas, beliefs, opinions, and knowledge in human groups – more generally. This literature, which ranges across the social sciences and philosophy, has yielded many important insights about human knowledge production. Only those insights especially germane to thinking about scientific communities will be covered here.[1]

The different sections of this Element will mostly be organized around different modeling approaches. Section 2, The Credit Economy, looks at models where scientists seek academic credit. These models are derived from game- and decision-theoretic approaches that treat humans as utility maximizers in order to explain and predict human behavior. As we will see, this work on the material incentives that scientists face yields insights on topics ranging from the division of scientific labor to sharing of academic research, to fraud. Section 3, The Natural Selection of Science, looks at models with a slightly different assumption – various people and practices in scientific communities undergo variations of selective processes similar to those seen in biological populations. By focusing on selective processes, these models elucidate emergent phenomena that go beyond the credit-seeking choices of individual scientists. These include the persistence of poor research methods, the effects of interdisciplinarity on progress, and industry influence on science via strategic funding. In Section 4, Social Networks and Scientific Knowledge, we see models that focus on the social connections between scientists and consider how these social connections impact things like theory change and belief spread. As will become clear, the way information flows in scientific communities deeply impacts the progress of science. Section 5, Epistemic Landscapes, focuses on problem choice in science and how strategies for problem choice can benefit or slow discovery. In particular, this section considers "landscape models" that represent a space of problems scientists can move through and explore. These models shed light, in particular, on the role of cognitive diversity in scientific communities.

---

[1] Readers may also notice that this Element puts extra focus on the literature coming out of philosophy of science. Throughout, I try to incorporate a broad range of work modeling scientific communities. But I am a philosopher after all.

Section 6, The Replication Crisis and Methodological Reform, considers a set of models with formal similarities that are also topically unified. This is because the models in question were developed alongside the metascience movement in response to the replication crisis. For the most part, these models consider various statistical practices and how they impact data gathering and inference in scientific communities. Central questions include: Why have so many findings failed to replicate? What are the main incentives and practices leading to this failure? What responses and interventions can improve scientific practice in the future? The conclusion of the Element summarizes and synthesizes policy recommendations from models throughout the Element.

Before beginning, I want to take a little more space for a (brief) discussion of model epistemology. What can models like those presented here tell us about science? And how should we take them to inform our understandings of it? Science is complicated and multifaceted. Science is diverse and varied. Any attempt to yield general theories of the workings of "science" will necessarily fail. The models presented in this Element are in keeping with an approach that works piecewise to improve the understanding of certain features, processes, and parts of the scientific enterprise. As such, none of the modeling results here should be taken as the be-all-end-all on some topic. Rather they are just one set of investigations contributing to our understanding of this complex and long-standing human enterprise.

Even so, we need to be careful about what we take away. Whenever simplified models are used to study complex social realities, there is room for error. Sometimes models fail to account for important real-world factors and, thus, support mistaken conclusions. Sometimes models abstract away from their target systems so severely that it is hard to assess their value.[2] That said, as we will see in this Element, models can nonetheless play a variety of important roles in the study of science. They can suggest new hypotheses for future study, challenge impossibility claims, suggest interventions that might not have been obvious, identify ways that proposed interventions might go wrong, and so on. In addition, they can act as an aid to ordinary reasoning or theorizing. Whether any particular model will be appropriate for some epistemic role will depend on the details of that model and the ways it is used. Many of the models presented here can play successful roles in some sorts of argumentation and inference,

---

[2] I do not really think that models are importantly different from many empirical investigations in this regard. Most studies involve abstracted representations of a full reality (say, twenty subjects answering questions in a lab). Like models, empirical data can only support inferences about the world that are appropriately tuned to the data gathered and the system it targets. That said, with highly simplified social models, there are often many ways that such inferences can go wrong.

even if they are not appropriate for others. While I will not be able to assess the quality and applicability of every model discussed, I will often describe the ways I think they are successfully used in argumentation about science.

As an example, many of the models overviewed focus on assessing various policy proposals. They are useful tools for doing so because it can be costly and/or difficult to implement new social policies. Models are a relatively cheap and easy way to start exploring how some new policy might impact practice. But it is risky to go directly from modeling outcomes to policy proposals for the reasons just mentioned. Instead, some of the models presented generate new (sometimes unexpected) hypotheses about what sorts of outcomes can follow policy interventions. When they do, it is often worthwhile empirically testing these hypotheses. In this way, a simple model does not tell us directly about what will happen in a complex reality, but opens up possibilities for study. While the model in question would not be an appropriate tool to directly shape policy, it is an appropriate tool to spur further exploration.

This example includes a generalizable lesson. The models described in this Element sometimes yield take-aways that are straightforward. More often, the best way to employ them to understand and improve science is in combination with empirical methods and other sorts of theorizing. Empirical studies of science help us build good models. Models help shape theory. Theory directs empirical research, which sometimes prompts further modeling. Via this sort of back and forth, models and empirical tools can work hand-in-glove to improve our understanding of the complex processes underway in scientific communities and help us shape the future of science.

2 The Credit Economy

Zihan works in astrophysics and had been planning to investigate a certain, exciting pattern in nebula formation. When she hears that another very prominent team is working on the same problem, though, she worries that they will publish first and get credit for the discovery. She decides to switch her group to another more modest project.

Jerome studies emotions in infants. After preparing his latest manuscript, he spends a long time deciding which journal to send it to. The most prominent ones would really boost his reputation but take a long time to review. The chance of his work getting accepted is low, and he might waste his time submitting. In the end, he tries for a lower level journal since his tenure review is coming up the following year.

In graduate school, Firuzeh's mentors were highly critical of her work unless it was absolutely stellar. What she did not know was that these critical reactions were shaped, at least in part, by the fact that she was a Muslim woman. Over time, she developed an expectation that finished academic work should be of extremely high quality if she wanted to get it accepted for publication, and, as a result, she started taking a very long time to perfect her work before submission.

Alice and Andy are two co-PIs working on human gene sequencing in competition with a number of other labs. They develop a new technique that will allow them to yield gene sequences much more quickly. If they share this technique, they will be credited for having discovered it, but other labs will be able to use it. If they wait, they risk another team developing the same technique and getting credit for it. But in the meantime, their research will go more quickly than their competitors. In the end, they decide to wait to share their new technique until they are further along in the project.

• • •

All the aforementioned examples involve scientists who are making decisions for strategic reasons. The branches of mathematics typically used to model this sort of decision-making are game and decision theory. In such models, agents are usually treated as utility maximizers. These models assume agents' actions yield different payoffs depending on either the behaviors of interactive partners or the structure of the world. By supposing that agents will prefer whatever actions maximize their expected payoffs, the models help explain and predict strategic behavior in humans.

Credit-economy models of science apply game- and decision-theoretic models to science but with a twist. Instead of maximizing payoff generally, these models assume that scientists attempt to maximize "credit." While credit in this sense is not a perfectly defined concept, it approximately tracks reputation and status in science, and attending benefits: fancy jobs, good pay, prestigious talk invitations, and so on.[3] In each of the aforementioned examples, the scientists in question made decisions not because they wanted to increase their production of useful knowledge but because they wanted successful careers. The sociologist of science Robert Merton was one of the first to clearly describe the credit motives of scientists (Merton, 1973). In his, and subsequent, work, it has been well established that many of the decisions scientists make day to day are indeed driven by credit motives. These motives, in turn, are shaped by credit structures of science – norms like the "priority rule," which gives credit

---

[3] Dasgupta and David (1994) describe the credit system as follows: "the greater the [scientific] achievement, the larger the rewards – which may come eventually, if not immediately, in the form of salary increases, subsequent research grants, scientific prizes, eponymy, and, most generally, peer-group esteem" (499).

only to the first scientist to make a discovery, journal practices, like publication bias (only publishing positive results), grant-giving rules, and so on.

In this section of the Element, we consider models that start with the assumption that scientists are motivated by credit and see how these motives might shape outcomes in science. There are a significant number of discussions that describe and defend this general approach. We will not address these arguments in detail, but interested readers should see Dasgupta and Maskin (1987), Goldman and Shaked (1991), Dasgupta and David (1994), Stephan (1996), Polanyi et al. (2000), Leonard (2002), Hull (1988), Strevens (2011), and Zollman (2018).

2.1 The Division of Scientific Labor

One might instinctively think of credit motives as a bad thing in science. Should we not expect that greedy or impure motives will drive scientists toward bad practices? And won't scientists with "purer" motivations, related to finding the truth, do better work? These questions actually go as far back as Du Bois (1898) and drive much of the literature described in this section of the Element.[4]

In an early credit-economy model, Kitcher (1990) argues that credit incentives can actually help scientific progress by improving the division of labor. It is typically desirable for members of a scientific community to work on an array of different topics or approaches. By doing so, they ensure that important discoveries are not missed. A community that is too uniform with respect to problem choice/theory adoption risks settling on theories that are suboptimal or failing to make potential breakthroughs. This is sometimes referred to as the "division of scientific labor."[5] But suppose that all scientists are purely motivated by a desire to discover true things. And suppose further that they have access to the same sorts of information and evidence. If so, they may agree on what topics for exploration are most promising and fail to divide labor effectively.

In Kitcher's model, scientists choose between projects, each with some intrinsic quality or tendency to succeed. He assumes scientists share an objective assessment of which projects are most promising. Thus, if they choose a project based on epistemic merit alone, they fail to divide labor. When scientists are motivated by credit, however, they are attracted to projects that fewer

---

[4] Du Bois (1898) argues for epistemic motives. In an even earlier work, Adam Smith argues that mathematicians and natural philosophers, unlike poets and fine writers, are not subject to credit-type motives and takes this to be a good thing (1759, part III, chapter 2). Many others have weighed in on this general debate, but we do not overview this literature for space reasons.

[5] Later, in Section 4, we will see a similar topic glossed as "transient diversity of practice" in science.

peers are currently working on. This is because they are more likely to be the one to make important discoveries on such projects and, thus, to receive credit (like Zihan who decided not to work on nebula formation). As Kitcher argues, "The very factors that are frequently thought of as interfering with the rational pursuit of science – the thirst for fame and fortune, for example – might actually play a constructive role in our community epistemic project" (16).[6]

Strevens (2003) uses a similar model to argue for the benefits of a specific credit-allocation rule in science – the priority rule. As noted, this rule specifies that only the first scientist to make a discovery receives credit, even if another scientist is unaware of the previous finding and even if the discoveries are nearly concurrent (Merton, 1957; Strevens, 2003). In Strevens's model, researchers again choose between projects and receive credit incentives either in line with (1) the priority rule, or some alternative, including rules that (2) give credit based on marginal contributions to research and (3) give credit to all scientists who make a discovery. He shows that all these incentive schemes can drive the division of labor but that, in doing so, the priority rule puts extra incentive on the most promising projects.[7] Strevens takes this result to help explain why science has adopted the priority rule. In science, a discovery need only be made once for its benefits to be conferred on society. In such a scheme, the division of labor yielded by the priority rule is particularly efficient on his model.[8]

Some have shed doubt on the usefulness of these models. Zollman (2018) points out that if scientists are motivated by a pure desire that the truth be discovered, they are already incentivized to divide labor in the ideal way to facilitate this discovery. Division of labor is the best way to ensure this discovery after all. Credit will only help if they are only motivated by a desire to discover the truth *themselves* and do not want another researcher to make the discovery. (But, one might ask, why would a truly truth-motivated scientist care who makes a discovery?) Bedessem (2019) argues that these and other models representing division of labor in science fail to track the complexity and variability of scientific problems/theories. Reijula and Kuorikoski (2019) criticize Strevens's model by pointing out that he fails to provide a mechanism for how credit incentives might emerge to effectively divide scientific labor. Goldman (1999) and Viola (2015) point out that there may be better ways to

---

6 Remarkably, in 1879, C. S. Peirce developed a model of division of labor in science with many similarities to Kitcher's. He was not interested in credit motives, though (Peirce, 1967).

7 See also Kleinberg and Oren (2011).

8 Strevens (2013), though, points out that if scientists tend to overestimate their likelihood to contribute to a research program, the priority rule in his model will drive too many of them to work on the more promising project.

coordinate division of labor that take into account centralized funding bodies (a topic we will return to in Sections 4 and 5). Muldoon and Weisberg (2011) criticize the assumptions that (1) scientists know how other scientists are distributing their labor and (2) can calculate the likelihood of success for different projects (thus calculating how they should best distribute their own labor). They develop an agent-based version of the model and find that when agents know only the research choices of a few community members, credit incentives do not work to divide labor. This is because agents do not have the proper information to incentivize them to choose the less promising alternative. And De Langhe (2014) points out that these models focus on dividing labor between existing options, rather than the exploration of new possibilities in science. He develops a credit model where agents can either explore new theories or test existing ones. He argues that the priority rule incentivizes exploration, while the fact that scientists tend to credit those working on similar topics to themselves incentivizes the study of existing theories. This addresses a different sort of division of labor in science – between exploiting the known and exploring the unknown. We will return to this issue at greater length when looking at epistemic landscape models in Section 5.[9]

In the end, do the models support the claim that credit incentives improve scientific division of labor? The evidence is mixed. A further observation is that scientists generally are complex and different. They do not typically assess the potential quality of theories or research topics in similar ways. They have different training and different interests that shape their research choices. Division of labor in science is often driven by these sorts of factors as much as credit motives. In assessing whether credit incentives in science are beneficial, and how we should shape them, there are other, arguably more pressing, issues than division of labor, which we turn to now.

2.2 Replication

Romero (2017) points out that Strevens (2003), and others advocating the benefits of credit motivations for division of labor, fail to consider the importance of replication. A hallmark of scientific knowledge is that it is replicable – re-running a scientific test should generate the same outcome. But the "replication crisis" has created massive upheaval as researchers in a number of disciplines have discovered that many core findings fail to replicate (Begley and Ellis, 2012; Open Science et al., 2015; Baker, 2016). (We return to this

---

[9] In contention with the models from Kitcher and Strevens, see also Dasgupta and Maskin (1987) who argue that credit incentives will instead push scientists to herd onto the same problems and approaches.

issue at length in Section 6.) As a result, many have advocated for researchers to spend more time replicating extant results. The priority rule, though, strongly disincentivizes replications, by assigning credit only to new discoveries. In other words, when we look at yet another sort of division of scientific labor – between seeking out new findings and verifying old ones – the priority rule causes problems. In support of this claim, Higginson and Munafò (2016) develop a model showing how the priority rule will tend to disincentivize the replication of existing results in favor of novelty, for just the reasons described.

In response to these sorts of issues, several authors argue that scientific communities should shape credit incentives to directly promote replication. Begley and Ellis (2012) argue that replications should always be required alongside new findings in order to publish. Romero (2018, 2020) advocates creating groups of scientists whose careers are entirely devoted to reproducing extant work. For these scientists, all credit then derives from attempting to replicate other experiments. On these proposals, credit incentives are re-engineered to avoid issues with priority.[10]

2.3 Fraud and Corner Cutting

Another worry about credit motivations is that they drive scientists to commit fraud or else to cut corners and engage in sloppy or imprecise research practices (Merton, 1973; Casadevall and Fang, 2012). Scientists who seek "fame and fortune" might be more likely to fabricate data supporting an impactful result. Likewise, scientists who aim to publish a lot of research quickly to gain credit, or win priority races, may be more likely to do sloppy work. Studies suggest that serious types of fraud are relatively rare but not insignificant in science. Most estimates put the percentage of researchers who have committed fraud at 1–3 percent (Fanelli, 2009; Bauchner et al., 2018; Xie et al., 2021) though some estimates are significantly higher (Gopalakrishna et al., 2022), especially those derived from reports estimating fraud among colleagues rather than oneself (Fanelli, 2009). In addition, estimates of the prevalence of less serious questionable research practices (QRPs) are much higher (Fanelli, 2009; Xie et al., 2021; Gopalakrishna et al., 2022).

---

[10] Lewandowsky and Oberauer (2020) use a simulation to explore another topic related to replication – the costs and benefits of either requiring replication of studies before publication or replicating only "high interest" (i.e., highly cited) papers after publication. The former has been advocated as a way to prevent the spread of false claims. They argue that both approaches are equally successful and advocate for the latter as more efficient in that fewer replications are required. However, their model does not consider practical costs to holding temporary false beliefs or the difficulties of retracting false research. Thus, the efficiency benefits they outline may be outweighed by other concerns.

Zollman (2023) presents a model showing how pernicious this problem might be. Scientists choose between fraud and honest research and, in doing so, attempt to maximize expected credit. His analysis shows that fraud can pay even when scientists are punished if caught. Lucky fraudsters – those who do not get caught for a significant period of time – yield higher credit payoffs than any other individuals. Heesen (2021) further shows how fraud early on can set a scientist's career on a more successful track. Because of the way that credit advantages tend to accumulate in science (see Section 2.6), fraud can pay even in cases where it is strongly disincentivized by punishment. Both authors suggest that these analyses may help explain the presence of fraud in research communities.

Short of fraud, both Higginson and Munafò (2016) and Heesen (2018) present models showing why credit motives might lead to the publication of swift and sloppy work, thus exacerbating issues with replicability. On the simple (realistic) assumption that the speed of research trades off with quality, credit-motivated actors tend to choose speed in order to increase their publication numbers. These choices, in turn, harm the quality of published research.

In light of these problems, Bright (2017b) gives a surprising defense of credit motives by developing a model to show how purely epistemic motives can also incentivize fraud. Agents in his model test questions in the world and derive sets of random, but truth-conducive, data, which they may then publish. As he shows, epistemically motivated agents are incentivized to misrepresent their data in cases where they believe that data is misleading. They are tempted to commit fraud for the good of truth just as credit seekers are tempted to do so for their own good. Huebner and Bright (2020) and Bright (2021) consider the various impacts of credit motives on fraud, though, and ultimately conclude that credit motives have indeed driven the adoption and spread of fraud (and other poor research practices). Given the clear harms of shoddy research practices, this raises worries about credit incentives.

Bruner (2013) presents a game-theoretic model investigating the sorts of incentives that might promote "peer-policing" of fraud in science. He gives his scientists the options to honestly report the importance of their work, to cheat by reporting greater importance, or to police by reporting honestly and also paying a cost to detect cheating by others. As he shows, providing credit incentives for scientists to spend time detecting fraud can increase the proportion of honest scientists in the community.[11] In this way, a new type of credit incentive might be able to alleviate problems with classic credit incentives.

---

[11] Bruner's paper actually provides an evolutionary game-theoretic analysis, which assumes that scientists learn over time which behaviors will benefit them. This analysis bears some similarity

2.4 Communism

While the priority rule may drive fraud and sloppy work, it plays another very important, positive role many have highlighted – promoting the swift, free sharing of scientific findings (Dasgupta and David, 1994). Merton (1942) describes this as the "communist norm." What is produced is also shared. The priority rule ensures that scientists are strongly incentivized to share their research findings as soon as possible. Without this incentive, scientists might instead prefer to keep the practical benefits of their research for themselves or to sell them to the highest bidder. Given that science progresses through the shared work of thousands of researchers and that it is only through sharing that critique of research and development of theory can happen, the communist norm is absolutely crucial to the workings of modern science.

But both Dasgupta and David (1994) and, more extensively, Strevens (2017) use models to argue that this sharing incentive will not always work. Paradoxically, the priority rule should also incentivize researchers to hide interim discoveries that can contribute to larger findings. Until the moment a discovery becomes publishable, the priority rule actually disincentivizes sharing. This is especially true during priority races where multiple teams are working on the same problem (as with Alice and Andy's gene sequencing in the introduction to this section). Strevens draws on cases to illuminate this point, such as the race to construct the TEA laser, as outlined in Collins (1974). In that case, "rivalry among competing labs discourage[d] communication that might have sped everyone to their joint goal" (Strevens, 2017, 3). For this reason, intermediate sharing can be understood as something like a prisoner's dilemma, where each individual wants to hear about others' findings but does not want to share theirs. On this model, Strevens characterizes the communist norm as a social contract – scientists adhere in order to create a system where all benefit, even though each scientist would prefer to defect by not sharing. Dasgupta and David (1994), on the other hand, argue that because of the benefits of reciprocity in the prisoner's dilemma, scientists will form mutually beneficial groups to share information even without some explicit social contract. On this picture, sharing will emerge naturally, though there will still be constant temptation to cheat by, for example, hiding detailed craft knowledge relevant to producing new results or being more willing to share the trade-secrets of other labs than one's own.

Both Heesen (2017b) (building off an earlier model by Boyer [2014]) and Banerjee et al. (2014) complicate this characterization. Heesen's model

---

to the selection-type models in Section 3, but, at heart, the results are driven by facts about credit incentives so they are included here.

involves multiple stages of publishable research, where scientists can gain credit by sharing at each stage. If they withhold information, they risk losing credit for that stage to another team, though they also increase their own chances of finishing the next stage first. Heesen shows that the expected equilibrium of this game is for all scientists to share, unless intermediate stages do not receive enough credit. Banerjee et al. (2014) have a model where teams work on various multistage solutions to a problem, and show that rewarding credit proportional to the subtasks they perform incentivizes intermediate sharing. On this picture, the solution to problems with communism is not to create a social contract, or depend on reciprocity, but to ensure that adequate credit is offered for all intermediate research advances that benefit the community to learn about. Note that this may not solve the issue raised by Dasgupta and David (1994), where scientists may withhold not the actual results but useful knowledge about how to produce further results. This issue might suggest the implementation of stronger incentives around sharing novel research techniques and methods.

Boyer-Kassem and Imbert (2015) also use a stages-of-research type model to address another phenomenon – the mushrooming of academic collaboration. This has been explained via appeal to epistemic synergy – two heads are better than one (Thagard, 2006). In their model, only the final stage of research receives credit. They find that even without synergistic effects, collaborative teams tend to pass through stages more quickly since the chances that some team member completes a stage are higher than that for an individual. This saves time and also increases their chances of receiving credit. Thus, in cases of priority races collaboration can increase local, intermediate communism by bringing the credit interests of a group of researchers in line. Collaboration promotes sharing within the group, even when there might be credit incentives not to share between groups.

We have been assuming here that communism is an unmitigated good in science, though there are a few models that indicate this might not be so. As we will see in Section 4, sometimes too much information sharing in a community can decrease diversity of exploration in a harmful way. By publishing an intermediate result, a team might prevent other groups from exploring alternate pathways to a solution that ultimately might be more successful. In addition, Bergstrom et al. (2016) develop a model with possibilities for intermediate sharing but where scientists can choose to opt out of a problem. (In this way, they address both communism and the division of labor in the same framework.) They argue that communism can impact the division of labor since intermediate sharing may prompt other scientists to abandon an area of research, thus potentially slowing progress to further discovery. Thus, whether or not intermediate sharing increases or decreases the speed of discovery may depend on countervailing forces – the benefits of catching many teams up to an intermediate finding and the detriments of scaring credit-motivated researchers off an important problem.

2.5 Credit and Productivity

One other factor weighing on the side of credit motives relates to scientific productivity. Dasgupta and David (1994) give an argument that, besides communism, the main benefit of the priority rule in science is to encourage the speed of scientific findings. As they point out, in a priority race, scientists will dedicate a great deal of time and effort to their work to ensure they finish first. Stephan (1996) formalizes this in a model where she shows how the priority rule incentivizes scientists to produce and share research.

Zollman (2018) extends this argument to credit incentives more generally. In his model, scientists in a community can choose to allocate time to research or to leisure. As he shows, scientists who are purely truth motivated – in the sense that their goal is for someone to discover truth – allocate a suboptimal amount of time to research even on their own lights. This is because they must give up a private good (leisure time) to produce a public good (research), which all benefit from equally. Credit motives help create private incentives for them to allocate more effort to research, and thus improve the public good. Generally, these last two upshots of credit incentives – the promotion of communism and productivity – are crucial to science and may be hard to generate via different incentive structures. These, more than division of labor, are arguably the features of current credit incentives most important to protect.

2.6 The Matthew Effect

One worry about credit incentives, which goes beyond their impact on scientific functioning, is that they are rewarded unfairly. Merton (1968) describes what he calls "The Matthew Effect" thus, "eminent scientists get disproportionately great credit for their contributions to science while relatively unknown scientists tend to get disproportionately little credit" (57). Work by Merton and others has demonstrated that such a pattern does, indeed, tend to hold in science.[12] Merton explains it as a negative effect of a reasonable tendency to pay attention to the work of renowned scientists (because it is presumably more likely to be important).

---

[12] This effect is named from the book of Matthew in the Bible – as Billie Holiday sings "them that's got shall get."

Strevens (2006) argues that the Matthew effect is not actually unfair. On his picture, credit is (and should be) allocated based on positive impact on society. Preeminent scientists are considered more trustworthy, meaning their findings are more widely influential. Because they are, in fact, more influential, they are thus deserving of more credit. He even argues that the Matthew effect might benefit society by attracting the strongest scientists to the most important problems.[13] Kleinberg and Oren (2011) develop a model, drawing from Kitcher (1990), that similarly argues for the benefits of a Matthew effect on the division of labor. In their model, some scientists get more credit than others. When those scientists choose promising projects to work on, their presence drives colleagues to select other topics because they, in turn, expect relatively low credit for competing with a prominent team. Thus, the Matthew effect acts like a symmetry breaker determining who will work on the best topics.

Heesen (2017a) presents a much less optimistic analysis of the Matthew effect. He develops a model where academics choose which papers to read based on epistemic quality. The patterns of attention that emerge tend to be highly stratified, with far more attention paid to the best work. Heesen points out that if general competence of an academic is highly correlated with the quality of the work produced, this pattern will tend to direct attention to the most competent academics and their valuable future work. But if findings are significantly dependent on luck, there is no future benefit to this stratified attention. The Matthew effect will then draw attention to future work of lucky academics but, in doing so, will not fairly reward them for benefits to society or improve scientific progress.[14]

Relatedly, Rubin and Schneider (2021) are concerned with the impacts of scientific network structure on priority and credit-giving. As they point out, credit is not automatically gifted by a community as a whole at the moment of discovery but is often established piecemeal via individual credit-giving. (This is part of the reason that intense priority disputes arise in science, where multiple academics claim to have priority to the same discovery [Merton, 1957].) In their model, scientists are connected in a network representing their social and communicative ties. Two scientists can make a discovery at the same time. Subsequently, other scientists give credit for the discovery to whichever scientist is closest in the network (or whichever has been cited closest to them). This

---

[13] I do not explain why this is supposed to happen because I do not fully understand this part of Strevens's argument.

[14] See also Watts and Gilbert (2011) who use an agent-based model to argue that a Matthew effect in citations does not improve the ability of a group to find good solutions to problems. They look at NK landscape problems, which we will address later on.

tracks a reasonable assumption – scientists often learn about discoveries from peers and social connections. But it also means that one scientist can end up receiving the lion's share of credit for a concurrent discovery due to network effects. As they point out, scientific communication networks often follow a "power law" type distribution – where a few scientists have many connections and many have fewer. They assume that their network has these properties and that, in particular, older scientists tend to be better connected than new ones. Under this assumption, credit allocation tends to unfairly benefit older, more connected scientists, generating a Matthew effect.

Rubin (2022) provides an even more dramatic demonstration of this sort of runaway process. In her network model, different researchers have opinions of the reputations of others, represented by weighted edges. As work is shared, it increases the weights directed toward a researcher, but more weights also increase the chances that others share her work. In this model, prominent researchers become more prominent, regardless of the reasons for their initial prominence. In such a process, clearly, there is no social benefit to the Matthew effect.

We might also be interested in credit inequity of a different sort. It has been noted in a number of disciplines that women tend to be cited less than men.[15] Rossiter (1993) uses historical evidence to forcibly make the point that women, in general, receive less credit than men for the same work. She dubs this the "Matilda" effect. In scientific disciplines where women and people of color have traditionally been excluded, they will tend to be newer to various professions and, thus, less connected. In the models just described from Rubin (2022) and Rubin and Schneider (2021), this will mean a credit disadvantage. In addition, Rubin and Schneider point out that homophily, or disproportionate in-group connection, can exacerbate this effect. They build models where scientists, realistically, tend to connect with those in their own identity group with higher probability. As they show, this can lead to persistent credit disadvantage for minority groups, who, due to homophily and relatively small numbers, tend to be less well-networked.

Thus, the set of models described here show how the Matthew effect can operate on a number of discrepancies – quality of work, age, demographic marginalization – or on just pure chance. In general, these models lend weight to the worry that the Matthew effect can lead to unfair credit-giving in science without creating a benefit for scientific progress or society.

---

[15] See, for example, Ferber and Brün (2011) and Dion et al. (2018).

2.7 Peer Review, Journals, and Funding

A core feature of science is the peer review process run by grant-giving agencies and academic journals. Colleagues assess the quality of research (either planned or completed), offer suggestions for improvements, and make decisions about funding and publication. By dint of shaping these decisions, the peer review process is one of the institutions in science that most directly influences credit incentives. A number of models consider how the incentives generated by peer review impact science.

Gross and Bergstrom (2019) worry about massive inefficiencies in the grant review process. Academics spend significant amounts of research time writing grants (Link et al., 2008). Given that many scientists cannot publish (and thus receive credit) without a grant, they are strongly incentivized to allocate time to grant-writing, even though it often takes time away from other scientific work. Gross and Bergstrom draw on contest theory to develop a model where academics make costly investments in attempts to win a prize (funding). Unsurprisingly, agents in their model are willing to pay high costs to gain funding, creating inefficiencies. They use their model to show that when grant funding is done via a partial lottery that weeds out poor projects and then randomly funds the rest, efficiency improves. Actors are incentivized to spend just enough time grant-writing to prove their projects are meritorious, rather than devoting more significant effort to beat out other scientists for funding. In developing this model, they show how specific details of funding structures can incentivize credit-motivated scientists to spend more or less time on research. We will return to potential benefits from lottery funding in later sections.

Nosek and Bar-Anan (2012) and Heesen and Bright (2020) likewise argue that prepublication peer review for journals creates inefficiencies in science. Peer review is time consuming and is often performed multiple times (in secret) on the same papers. Authors spend time on multiple revisions, which may or may not improve the quality of their work. Given current credit motives, scientists are willing to engage in this inefficient process because it is the only avenue to credit. (Simply posting on preprint servers, for example, spreads results but does not generate the same sort of prestige or support promotion.) On the basis of this analysis, these authors suggest abolishing prepublication peer review altogether. On this picture, peer review would occur after finished articles are posted on preprint servers. Presumably the most important articles would receive the most criticism and feedback. Arvan et al. (2020) use a model drawing on the Condorcet jury theorem to support this proposal. They point out that postpublication peer review, when it works well, can take advantage of the opinions and insights of many scholars, rather than just a few reviewers.

Assuming all these scholars have decent judgment, more judgments are better in determining the ultimate importance of the work.

Rubin (2022) raises worries, though. She uses network models (like those described in Section 2.6) to argue that postpublication review, since it is necessarily de-anonymized, may follow popularity rather than merit. This could have negative impacts, as noted, for young and marginalized scholars. Arvan et al. (2020) point out that short-term anonymized postpublication review is possible, which may partially mitigate these worries. Though if scholars are worried about reciprocation from powerful peers, eventual de-anonymization will still likely impact their willingness to share judgments in reviews.[16]

Tiokhin et al. (2021), drawing on work from economics, use game-theoretic signaling models to suggest other solutions to the inefficiency of peer review. As they point out, authors are incentivized to submit their papers to high-quality journals in hopes of receiving high levels of credit, even when their work is not of high quality. This leads to a lot of extra reviewing work sifting through papers of various qualities and rereviewing when authors must submit work multiple times. It would be more efficient if authors tailored their submission choices appropriately. They point out that in signaling theory, costs can support honesty. In this case, authors may only be willing to pay a cost to submit if they anticipate a high enough likelihood of being accepted (like Jerome who decided not to pay the cost of a long review period at a top journal). This should deter lower-quality submissions. They suggest various possible costs including direct ones but also costs just for resubmitting, or limits to submission attempts.[17] Along these lines Azar (2005), Leslie (2005), and Cotton (2013) all argue that slow review processes, which act as a cost of submission, can deter nuisance submissions.[18]

Note that these results seem to raise a worry for the proposals described earlier by Gross and Bergstrom (2019). If submission is faster and easier, maybe academics will be incentivized to submit too many poor grant applications. But, on the other hand, their proposal would also significantly decrease the costs for review, meaning that perhaps extra submissions would not translate to too much extra work for reviewers.

Moving away from analyzing efficiency costs, Gross and Bergstrom (2021) are interested in the impacts of reviewing on problem choice and especially on

---

[16] Neither of these papers look at credit incentives exactly. They are included here for their relevance to the debate about postpublication peer review.

[17] Tiokhin et al. (2021) make a somewhat related observation about the benefits of high start-up costs for research projects. These costs incentivize researchers to spend more time on each project; more in Section 3.

[18] See also Oster (1980) who gives an early credit analysis of journal submission choices.

risk taking. Their model compares the impact of ex-ante review (of proposals for new work, i.e., by grant-giving agencies) with ex-post peer review (of finished work, i.e., by journal reviewers) on the choice of research topic.[19] They assume that researchers value projects that they predict will have a significant impact on beliefs.[20] As they argue, ex-ante review will tend to incentivize more conservative proposals. Suppose a scientist has a private reason to believe a risky or unusual experiment will succeed. He then has good reason to carry out this experiment, with the expectation that journal reviewers will be impressed with his impactful finding. But he will anticipate that grant-giving bodies will reject the same project for funding, as reviewers will not tend to anticipate its success ex-ante. Thus, ex-ante reviewing incentivizes a safer choice – one that more academics agree is likely to succeed. There has recently been some movement toward ex-ante review for not just grant proposals but also for publication via registered reports (we will return to this in Section 6) (Soderberg et al., 2021). Gross and Bergstrom's results suggest that such a move might decrease the production of high-risk, high-reward science. A number of academics have raised concerns about trends toward conservatism in research and the impacts of such trends on discovery (Currie, 2019; Stanford, 2019; Wu and O'Connor, 2023).

One journal practice that will come up extensively in Section 3 and 6 for its role in shaping incentives in science is publication bias. This refers to the practice of preferentially publishing novel, positive findings over null results (or replications). A key worry about publication bias is that it incentivizes scientists to obtain positive results regardless of what their data look like. In the context of current statistical standards, namely null hypothesis significance testing or NHST (more on this later), this creates pressure to use a number of QRPs to yield positive, significant findings. These QRPs, including p-hacking, HARKing, and forking paths, will be addressed at length in Section 6, which also looks at proposed incentives to mitigate their harm.

2.8 Incentives and Gender Identity

Generally, across academia, women tend to be less productive than men, in that they publish fewer papers per unit of time (Etzkowitz et al., 2008). Those who study science have attempted to explain this "gender productivity gap" via

---

[19] Note that this is not quite the same as prepublication and postpublication review. That draws a distinction between two possible periods of review for finished work. Gross and Bergstrom (2021) are interested in review of proposals for new, incompleted work versus review of finished work.

[20] Note that this is not exactly a credit motive but is consistent with either a desire for credit given to high impact research or an epistemic motivation to improve belief.

appeal to many factors – maybe women have less time than men due to family responsibilities, maybe they face discrimination in peer review, or maybe they are inherently less productive than men.[21]

Bright (2017a) and Hengel (2022) model a suggestion discussed in Sonnert and Holton (1996) and Lee (2016). The proposal is that women might be less productive because they expect to face higher standards for publication. On this suggestion, women have been socialized to expect more push back on their academic work and to compensate for this. In Bright's model, academics prefer to publish as many papers as possible, but women tend to believe they must do higher quality work in order to publish (like Firuzeh who learned from advisor expectations to overperform). On this assumption, he shows that women will tend to submit (and publish) papers less often, though each paper will represent a greater effort.[22] Hengel (2022) goes further in validating a decision-theoretic model using publication data from economics. For instance, she points out that if women are responding to this sort of incentive, there should be a quality gap between men's and women's writing that increases over time. She indeed finds such a gap using several measures of writing quality.[23]

In reality, multiple causal factors likely contribute to the gender publication gap. This modeling work carefully elucidates one of these.

● ● ●

Inasmuch as scientists seek credit for their work, credit-economy models can be used to assess benefits and detriments of various credit incentives in science. As we have seen, these models have been used extensively to present and assess policy proposals, especially those related to reworking credit incentives in science. In the conclusion of this Element, we will return to some of these. In the next section, we switch focus to ask how credit rewards shape not the individual choices of scientists but the selective processes of scientific communities.

3 The Natural Selection of Science

Dr. Alison Slacks developed a new, cutting-edge method to research virus evolution. As a result, her work has gotten a good deal of attention for its relevance to a worldwide pandemic. Her papers were published in *Science* and *Nature*,

---

[21] See Bright (2017a) for more on these possible explanations.

[22] As he points out, "The model thus brings to the surface the role of the 'publish or perish' norm in producing and perpetuating a productivity gap between men and women" (423).

[23] In addition, Alexander et al. (2021) find that women tend to spend longer revising papers, which is consistent with this explanation. On the other hand, they also find that less experienced referees take significantly longer to review women's papers, which may point toward the role of bias in the gendered publication gap.

she was asked to speak on Good Morning America, and was featured in *Time Magazine* and *Newsweek*. Her Twitter account has over 100k followers.

Dr. Slacks's colleague, Paul Pantalons, researches mechanical limbs. Although his work is just as meticulous and methodologically sound as Dr. Slacks's, it has not gotten the same sort of attention.

When their students have gone on the job market, Dr. Slacks's protégés have tended to have more success. Hiring committees have heard of Slacks's research program and are keen to hire students who might develop a similar profile. Plus, with all her invitations to speak at prestigious conferences, Dr. Slacks has been able to promote her students' work. The result is a number of newly minted researchers at top universities using her methods to study virus evolution. These students, in turn, influence their colleagues, give talks on their work, and start to churn out students of their own. The prevalence of this work leads colleagues to adopt similar methods. Work on virus evolution using Slacks's methods becomes increasingly prevalent. Meanwhile, many fewer people ever hear about Dr. Pantalons's mechanical limbs.

● ● ●

Evolutionary models can inform how selective processes, including processes of cultural selection, shape behavior. There are strong connections between these sorts of models and game-theoretic ones. Often the sorts of things that can be modeled using payoffs or utilities in game theory are the same sorts of things that contribute to fitness, and thus selection, in evolutionary models. In Section 2, we saw how models derived from game and decision theory could be used to study scientists as credit-seekers. Now we will see how models derived from evolutionary theory can inform how credit shapes selective processes in science.

In the example with Slacks and Pantalons, the hiring process in a discipline shaped its future practices and beliefs. In particular, because Dr. Slacks's methods and topic tended to accrue credit in the form of prestige and exposure, they also tended to persist and replicate in the community as her students sought jobs and colleagues chose new methods. This process was independent of how scientists were individually incentivized to choose topics or methods of study.

The idea that science evolves in ways analogous to natural selection is well explored and goes back to scholars like Popper (1972), Kuhn (1962), Campbell (1965), and Hull (1988).[24] As will become clear in this section of the Element, differential tendencies for people, methods, and theories to persist, or replicate,

---

[24] Hull (1988) especially develops lines of thought relevant here. He is mostly focused on the selective development of concepts or ideas in science, and accounts for social influences on this development.

may help explain various phenomena in science from low statistical standards, to conservative research choices, to the success of industry influence. Notably, the effects of these processes are sometimes orthogonal to the effects that incentives have on scientists's choices and are, thus, important to study in order to develop a fuller picture of the workings of science.[25]

3.1 The Natural Selection of Bad, Good, and Conservative Science

One important question is: how do selective forces influence the quality of methods used in science? Will selection tend to push scientists to use high-quality techniques? Fast, sloppy techniques? Or methods with some particular character?

Smaldino and McElreath (2016) present a model intended to explain the persistence of poor statistical standards in science.[26] As they point out, over many decades, there have been repeated calls across the sciences for statistical reform. But in spite of widespread awareness of problems like pervasively underpowered results, there has been little to no improvement in practice. Following many others, they identify publication bias as one culprit. As noted, many journals tend to accept new positive results and reject null results. This practice creates perverse credit incentives where researchers do their best to adopt whatever (potentially questionable) practices yield significant findings, regardless of whether those practices are truth-conducive (Nosek et al., 2012). Of course, one might respond that many researchers are disinclined to adopt poor research practices for ethical and epistemological reasons. As Smaldino and McElreath argue, though, poor methods can spread as a result of these incentives through dynamics that, "require no conscious strategizing – no deliberate cheating nor loafing – by scientists, only that publication is a principal factor for career advancement" (1).

Their model includes a community of N labs that each use characteristic methods. These methods are defined by a level of *efficacy* – the ability to positively identify true results – and *effort*, which is necessary to pick out and avoid false positives from this set.[27] Effort decreases productivity but improves the quality of research outputs by decreasing false positive findings. In each round of simulation, labs produce research with a characteristic rate and quality based on these factors. After this production of research, "evolution" occurs – one

---

[25] For another review of cultural evolutionary models in science, see Wu et al. (forthcoming).

[26] In doing so, they draw on their own earlier work in McElreath and Smaldino (2015). See also Smaldino (2019b).

[27] They use the term "power" instead of effort. As Stewart and Plotkin (2021) point out, this is easily confused with statistical power. For this reason, I use "effort" here.

lab is randomly chosen to shut down and is replaced with another. The new lab copies the characteristics of a chosen parent lab, where high-credit labs tend to be copied more frequently. In this way, labs that accrue credit tend to more strongly influence the community (just as Dr. Slacks had an outsized effect on her discipline compared to Dr. Pantalons).

Smaldino and McElreath find that this process tends to select for "bad science," that is, research methods that use minimal effort to avoid false positives. Instead, labs adopt methods that yield many (publishable) false findings. This is true even when labs are punished for publishing false results. In such cases, punishment disincentivizes individuals from choosing poor methods. A rational credit maximizer would actually choose better methods to avoid the chance of negative payoffs when their results fail to replicate. The labs that use poor methods and happen to never get caught accrue the most credit and are, thus, copied most often.[28] In this way, a selective process can lead to the proliferation of poor methods even when individual responses to credit incentives should push toward better ones.

Tiokhin et al. (2021) further support this finding. They use an evolutionary model to consider the impact of the priority rule specifically on research quality. In their model, multiple labs often end up in a priority race. They show that high-priority incentives drive the evolution of small sample sizes as labs compete to finish first, which increases the proportion of false discoveries.[29] They use the model to advocate "scoop protection" – an intervention employed by some journals to protect runners-up in priority races. With scoop protection, labs can still publish a finding that has been newly established elsewhere.[30,31] (Of course, one worry about this proposal is that credit is partially allocated via attention and citation by colleagues. If so, those who finish first may still get more credit, even if later papers are published.) These two papers together show how credit incentives can shape selective processes that lead to the spread of poor methods.

If selection leads poor methods to proliferate, is there a way to counter this process? Smaldino et al. (2019) address the potential for positive change. First

---

[28] Similarly, see Zollman (2023) and Heesen (2021) who develop models where the most credit accrues to lucky fraudsters who do not get caught and are, thus, in a position to have a disproportionate impact on the community.

[29] Note that this result dovetails with models from Section 2 on incentives for cutting corners and sloppiness.

[30] They also point out that high start-up costs for new projects can ameliorate this selective process. When scientists must pay to begin a new project, they are incentivized to spend more time on their current research and, thus, tend to gather more data.

[31] Tiokhin and Derex (2019) give an experimental verification of these results. Subjects under priority-type pressure tend to spend less time investigating a phenomenon before sharing their findings.

they argue that changes to the peer review process that counter publication bias, like open science and registered reports, can help. By creating venues for all high-quality results to be published, scientific communities decrease perverse incentives for false positive results (we will return to this idea in Section 6). Second, they add a funding stage to the model, where some labs receive grants that increase their productivity. These authors find that a hybrid lottery system – where funds are randomly allocated to a lab that surpasses a quality threshold – is effective at reducing the natural selection of bad science. This reward system leads to the "selection" of a wide swath of effective methods (rather than just those with the very highest credit, who often use questionable methods).

Stewart and Plotkin (2021) develop a similar model, but their aim is to show how incentives created by publication bias can actually drive selection for *good* science. They consider labs that have an additional option – to expend effort on theory before developing a hypothesis to test. Doing so increases the chances that the lab's hypotheses will be correct ones (and thus increases the chances of yielding positive results and receiving credit).[32] Their model has two stable endpoints: one where researchers use theory and effort to yield high-quality work, and another similar to that in Smaldino and McElreath (2016) where labs use minimal effort and publish many false positives. They show that the good science outcome is particularly likely when (1) replication efforts are used to punish labs with false findings and (2) the community avoids rewarding attention-grabbing (but likely to be false) findings. In this way, Stewart and Plotkin build on the work of Smaldino and McElreath (2016) to identify further interventions that might improve scientific practice.

We have seen how selection can impact methodological quality, but what about other aspects of methodology? O'Connor (2019a) develops a model to investigate a different worry about science – that researchers tend toward conservative choices by publishing many low-risk, low-reward papers.[33] In her model, labs choose between projects that generate small, consistent levels of credit or else large but inconsistent levels. While risky science generates less credit on average, the highest credit labs are those whose risk-taking paid off (much like low effort labs in Smaldino and McElreath [2016]). These labs have an out-sized effect on the community. One might think that high-risk, high-reward science should then spread via selection processes. But as O'Connor points out, risky science is often hard to copy. A successful risk-taking scientists must also get lucky to succeed, and their students may not be so lucky. In

---

[32] In Section 6, we will discuss the importance of good theory at greater length.

[33] While these papers generate dependable levels of credit, they may also be unlikely to yield truly novel or important discoveries (Luukkonen, 2012; Currie, 2019; Stanford, 2019; Schneider, 2021).

other words, its success levels are less heritable than for conservative science. When this is true, she finds that conservatism spreads via selection in science.

3.2 Self-Preferential Biases and False Paradigms

Another place selection can work is on the types of research frameworks scientists use. Akerlof and Michaillat (2018) ask whether selective processes in science can help explain the persistence of "false paradigms" – research frameworks that are less accurate or useful than some alternatives. As Kuhn (1962) famously argued, scientists tend to adhere to a paradigm even as anomalous evidence accumulates against it, in part for social reasons. In Akerlof and Michaillat's model, a community of agents uses paradigms of differing quality. Scientists in the model may be pre- or post-tenure. New scientists are trained by tenured scientists and tend to adopt their mentor's paradigm. These new pre-tenure scientists are then evaluated for tenure by a random member of the tenured community.

The decision to tenure a junior scientist is made on two grounds. First, tenured scientists may be more or less competent to distinguish between the quality of better and worse paradigms. Second, scientists may be more or less biased toward their own paradigm. (Empirical evidence suggests that in many ways scientists do exhibit this sort of self-preferential bias.[34]) They find that self-preferential biases can stabilize poor paradigms. Researchers tenure those who do the same sort of work they do, even though there is better work out there. This is especially worrying when researchers are not competent to tell which paradigms are of higher quality. The result is that the community remains stuck in the same, poor paradigm despite the presence of better alternatives.[35] They suggest this as a possible explanation of similar episodes in the history of science.

This raises the question of whether disciplines that employ poor paradigms are doomed. Akerloff and Michaillat point out that if competence improves, self-preferential biases may be overcome. Smaldino and O'Connor (2022) consider whether interdisciplinary contact might break this cycle. They look at a similar model but where (1) a scientist's work is sometimes assessed by someone outside their discipline, and (2) scientists sometimes copy methods from the other discipline. They show that a better paradigm can easily spread as a result of this sort of contact. Some individuals copy better paradigms. And those who

---

[34] See citations in Smaldino and O'Connor (2022).

[35] Heesen and Romeijn (2019) use a model to give a very different reason why current paradigms may be sticky. They point out that editors will tend to know more about currently popular research paradigms and topics. If editors try to maximize the quality of accepted papers, their decisions to publish will tend to favor these topics as a result of their knowledge imbalance.

adopt a high-quality, but rare, paradigm can become prominent and influential as a result of out-group credit-giving, leading to the spread of good methods in their home discipline. As Smaldino and O'Connor point out, this does not mean that we should eradicate a disciplinary structure altogether. An entirely flat disciplinary structure might be subject to the sort of entrenchment of poor methods that Akerlof and Michaillat (2018) identify. But, Smaldino and O'Connor argue that we *should* promote a decent level of interdisciplinary contact to facilitate the spread of good methods between disciplines. This proposal is supported by other, qualitative work identifying benefits of interdisciplinarity (and comes with few downsides).

3.3 Industrial Selection

Industrial interests often conflict with public belief, and when this is the case, industry has historically used a bestiary of subtle, sneaky, and effective strategies to confound belief both within and without the scientific community (Oreskes and Conway, 2011; O'Connor and Weatherall, 2019). Many have responded by proposing codes of good conduct for individual scientists (see, for example, Douglas et al. [2014]). But even if all scientists adhere to high standards, industry is still incentivized to shape scientific progress. What if industry actors can use selective forces to their own advantage?

Holman and Bruner (2017) use a selective model to show how industry funding can shape the course of scientific knowledge without ever corrupting the practices of individual scientists. In their model, a community attempts to identify which of two theories, A or B, is most promising. They do this by testing these theories. But while most scientists use truth-tracking methodologies, some use methods that, unknowingly, point toward the worse theory (or, perhaps, use methods that might otherwise be appropriate but because they are misapplied in the domain of interest are misleading). The model includes an industrial agent who can choose which scientists to fund, thus increasing their productivity. As they show, by funding just those scientists who use worse methods, the industrial agent can push the community toward a consensus that materially benefits industry. Findings supporting industry beliefs flood the community from well-funded, productive scientists using poor methods, thus shaping the future research of peers. This funding also tends to increase the prominence of researchers using poor methods, and thus the placement of their students, and the tendency of other researchers to copy them. This model demonstrates how, in principle, a focus on individual decision-making might lead policymakers to miss important negative impacts of industry on science.[36]

---

[36] On the basis of this research, Korf (2023) calls for a genuinely social/community level approach to thinking about preventing industrial influence in science.

3.4 Levels of Selection in Science

In biology, levels of selection theory considers how selective processes may act on different levels of organization. For instance, individuals derive payoffs, but the groups they belong to may likewise be thought of as having payoffs, or fitness levels.[37] Tiokhin et al. (2021) consider what would happen if scientific communities shifted the "level of selection" for scientific credit. In particular, they suggest that rather than rewarding credit to individual scientists, we might sometimes want to give credit to groups. (Of course, we already give credit to groups of collaborators, but the suggestion is that we might also try to reward other groups.)

As they point out, scientists directly influence the progress of knowledge through their own work, but they also, as members of tightly knit communities, have indirect effects on knowledge production via interactions with others. Scientists who produce careful peer review, give feedback to colleagues, take the time to share their code and data, do service work, mentor students, and so on can improve the research of many others. But these sorts of contributions take time and are not typically rewarded by hiring and promotion schemes. This creates incentives to shirk prosocial duties. They argue that researchers may even be incentivized to harm community members by, for example, taking advantage of student productivity or submitting overly negative reviews to delay publication by competitors. Furthermore, they argue that individual credit rewards disincentivize specialization into complementary roles, which typically improve the efficiency of production. By giving credit to groups, these perverse incentives may be reversed.[38] Their proposal has the potential to alleviate a number of issues raised in Section 2 – failures of communism at interim research stages, disincentives to perform replication work, and fraud. Although this is not a focus of their paper, it has also been pointed out that these sorts of community jobs are often disproportionately done by women and people of color in academia; so their proposal may, in addition, improve equity. The widespread move toward increasing collaboration in science already reflects a change that leads to credit-giving at a higher "level" than the individual. Rewards to well-run departments or high-functioning sub-fields may likewise help promote prosocial science.[39]

---

[37] There is a great deal of confusion and debate over levels of selection, but that is beyond the scope of this Element. See Okasha (2006) for a good overview.

[38] Relatedly, Dasgupta and David (1994) point out that there is little incentive for scientists to coordinate the timing of related research efficiently, and this proposal may help with that problem as well.

[39] To some degree, such rewards happen naturally – a well-functioning department might facilitate better research productivity all around and, thus, gain a better reputation. This might attract more funding and better graduate students, thus benefiting each member.

Worries remain, though. It is unlikely that in any realistic future, credit incentives in science will only be awarded at higher levels of selection. Thus, there will still be incentives to shirk prosocial duties if scientists can pursue credit individually while benefiting from the community credit generated by others. Note that this incentive structure looks like a public goods game, where incentives tend to push agents toward defecting on their own contributions. If so, rewarding higher levels in science might not have the positive effects on prosocial behaviors that these authors suggest.[40]

3.5 The Emergence of Discrimination in Science

A host of empirical results suggest that women tend to do more work per credit received than men, on average, when engaged in scientific collaboration.[41] How might such patterns of behavior emerge? And do they reflect selective forces in science?

Bruner and O'Connor (2017) and O'Connor and Bruner (2019) point out that coauthorship requires a bargain to determine (1) who will do how much labor, and (2) who will receive which authorship position. They model the cultural evolution of collaboration using a *bargaining game*, which involves two players who divide a resource. They find that two social groups often evolve systematically inequitable patterns of behavior, that is, where one group gets more credit, and the other less. In addition, they show how minority groups in science may be specially disadvantaged as a result of dynamics outlined in Bruner (2019).[42] As they argue, their models may help explain how patterns of collaborative behavior that systematically disadvantage groups like women may emerge.

Bruner and O'Connor (2017) also consider a different interpretation of this sort of model where the social identity groups track hierarchical levels in academia, like professors and graduate students. They find that power differentials can lead to systematic advantage for one group. If professors, for instance, are more powerful than graduate students, this can translate to better authorship positions and less work per credit earned. This may help explain phenomena

---

[40] Thanks to Jingyi Wu for this point.

[41] Feldon et al. (2017) looked at early PhD students in biology and found that while women put in significantly more work, men were 15 percent more likely to be granted authorship per hour worked. In economics, Sarsons (2017) finds that men's tenure chances get an equal boost from both coauthored and single authored publications. But while women's chances increase 8 percent on average for a single authored paper, they increase only 2 percent for a paper coauthored with a man. In other words, when collaborating with men, they are not granted equal credit for their contributions. In addition, across the sciences, women are less likely to hold prestigious authorship positions (Larivière et al., 2013; West et al., 2013).

[42] I do not go into detail here for space reasons. See also O'Connor (2017).

like "ghost authorship," where prominent academics demand author positions without actually contributing to a paper.

Both sets of results raise worries about fairness in collaboration and credit-giving. Since scientific norms emerge via cultural evolutionary forces, they are likely to follow the same sorts of patterns that other norms do. When it comes to bargaining, social identity markers often translate into inequity, and science may be no different.[43]

● ● ●

As we have seen in this section of the Element, credit influences not just individual decision-making in science but also selective processes. Which individuals get jobs and receive tenure? Which theories, paradigms, and methods persist and spread? And how should this inform policy? Considering these group-level processes helps to complicate and nuance our understanding of how credit impacts scientific functioning, and what we should do to improve progress. This is especially germane, given results showing that sometimes incentives push choice in one direction (i.e., toward quality research) but selection in another (toward questionable practices). In the next section, we continue to consider the impacts of extended, group-level processes on science but shift focus to look at processes surrounding communication specifically.

4 Social Networks and Scientific Knowledge

Jane Hightower was a physician who had been seeing an unusual cluster of symptoms in some of her patients – hair loss, nausea, fatigue, weakness, etc.. These patients tended to be health-conscious and wealthy. While driving home, one of Jane's colleagues heard a radio story about a town where many suffered similar symptoms after eating fish contaminated with mercury. After testing one of Jane's patients, the colleague discovered elevated mercury levels. As it turned out, the patient ate a lot of fish, especially those species that tend to accumulate the mercury released by coal power plants.

Dr. Hightower started to investigate the connection between seafood and mercury poisoning in her patients. She shared her worries about this link, and her growing evidence of it, with colleagues across the country. Some of these doctors began to watch for mercury poisoning in their own patients. She reported her findings to news agencies, and the show 20/20 ran a segment on her work. They tested for mercury levels in local grocery store fish, finding them elevated above FDA standards. As more evidence accumulated, several medical associations passed resolutions about the dangers of mercury. Government

---

[43] See O'Connor (2019b) for more on this sort of process.

agencies around the world eventually passed stricter policy guidelines to better regulate mercury levels in fish.[44]

• • •

Human knowledge is deeply social. People tend to adopt most of their beliefs through their social networks. We learn that the earth is round from a teacher. We hear about the benefits of prenatal DHA supplements from a friend. A newscaster reports a tsunami in Japan. This process of evidence and belief sharing is at the heart of human cultural advancement. Without it, we would not be able to accumulate the knowledge and expertise necessary for advanced technologies, medicine, or cutting-edge discoveries.

For this reason, many researchers study "epistemic network" models to learn more about human beliefs and knowledge. The nodes in these networks are individuals, and the edges are communicative links. New ideas, opinions, beliefs, or evidence can pass along these links and, thus, travel through a social network.

Scientists are no exception to the social knowledge rule. In the aforementioned (real-life) example, note that Hightower first got the idea of a mercury/seafood connection from a colleague, who, in turn, had heard about mercury poisoning and fish on a radio show. Hightower then spread this idea to colleagues and members of the media. Various colleagues changed their practices as a result of her information by beginning to test the mercury/seafood hypothesis themselves. They shared the new evidence they accumulated and helped alert more people to potential danger.

In this section, we consider what social network models can tell us about science. One primary focus is on how communication can shape exploratory tendencies in science. Other topics include polarization, industry funding in science, and the formation of scientific networks. And, as we will see, considering how information flows between scientists allows us to think further about how scientific communities work and how to make them work better.[45]

---

[44] This history comes from Hightower's personal account in Hightower (2011). See also O'Connor and Weatherall (2019) who use this case to illustrate the social nature of human knowledge production.

[45] Prepare for a tangent because I did not know where else to put this bit. Related to the question of how scientists share information and come to consensus is the question of how scientists *should* do so. Many researchers have used models to study how to rationally combine credences or judgments. Does this work inform how scientists ought to combine opinions? Solomon (2006) argues judgment aggregation (or voting) without deliberation may help reduce groupthink in science. Magnus (2013) thinks votes of this sort do not respect the importance of social values in determining when/whether we should endorse scientific findings. Bright et al. (2018) argue, though, that for scientific collaborations, majority voting on what claims to endorse in a paper does make sense because such a document need not reflect collective belief but rather is a different sort of communicative move.

4.1 Models of Evidence Sharing

There are different sorts of information/ideas/opinions/knowledge/beliefs that spread through human social networks. As such, there are a variety of network approaches to modeling the spread of doxastic (or knowledge-related) states in human groups. Most of the network models we will consider assume that agents share not opinions or testimony about their beliefs, but evidence. For instance, instead of modeling someone saying to a peer, "I think fish are causing mercury poisoning," these models better correspond to an individual describing data on how much fish her patients with hair loss are eating. This sort of model is especially germane to scientific communities where scientists produce evidence to support their beliefs and attempt to sway others by sharing it.[46]

The particular evidence-sharing models we focus on were first developed in economics by Bala and Goyal (1998) to study social learning. Agents test various actions or theories, gather evidence from their tests, and share that evidence with their network neighbors. Their neighbors, in turn, are influenced by evidence they encounter and use it to determine which actions or theories they should test and, thus, what sorts of data they pass on. Through this process, communities eventually tend to form a consensus, either correct or incorrect, about which action is best. One noteworthy feature of this model is that Bala and Goyal (and others using it) typically assume that agents learn using Bayes's rule, that is, they respond rationally to evidence. In cases where group learning fails, then, this is a result of emergent, communicative effects rather than individual failures of rationality. In this way, the modeling paradigm allows researchers to demonstrate something much harder to establish experimentally – that rational learners may fail in group settings. In addition, this assumption is well-tuned to scientific communities, where agents are experts trained to learn well from evidence.

Zollman (2007, 2010) first adopted this model to study theory change in science. Under this interpretation, agents may, for example, be trying to decide whether "cigarettes are safe" or "smoking causes cancer." Each scientist gathers and shares data guided by their preferred theory – those who think smoking might be dangerous actually test this possibility, while those who do not, ignore it. Or, in the Hightower example, the theories might be "mercury in fish is

---

[46] Two other dominant epistemic network paradigms are contagion/diffusion models and opinion dynamic models. The first sort assumes that ideas spread from person to person much like a virus can (Rogers, 2010). These are not used widely to model science, though LaCroix et al. (2021) use them to think about retraction and especially failures of retraction. Opinion dynamic models typically assume that agents hold an opinion between 0 and 1 and, in successive rounds, average these with neighbors (Hegselmann et al., 2002; Golub and Jackson, 2010, 2012). Grim et al. (2015) use this sort of model to argue that interdisciplinary links are important to science.

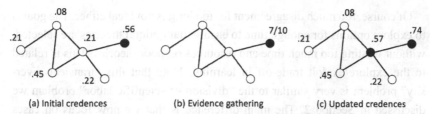

Figure 1 Example of an epistemic network model. Agents (a) have initial credences about the success rate of some action or theory, (b) those who prefer the action/theory actually try it, and (c) on the basis of their results, they, and connected neighbors, use Bayes's rule to update their credences

making patients sick" and "mercury in fish is currently at safe levels." Those adopting the first theory tend to gather evidence regarding mercury and fish, change their beliefs as a result of this evidence, and share it in ways that then impact current group consensus.

Figure 1 shows what a round of this model might look like. We see agents connected in a network. At the start of the round, (a), they hold initial beliefs or credences in some theories. Here we will follow Bala and Goyal (1998) in assuming that they are familiar with one action or theory A, which has a success rate of 0.5. The figure shows their credences about whether theory B is more successful than A. Those with credences less than 0.5 (represented by white nodes) prefer theory A. In step (b), the agents who prefer theory B will test it. Here just one agent tries such a test ten times and yields seven successes. In the final stage (c), the agents who observe this evidence use Bayes's rule to update their credences. In this case, two agents do so, and the evidence gathered convinces one agent to change their preference to theory B. The model proceeds round after round until all agents' preferences for theories are stable.

4.2 The Benefits of Transient Diversity

One might think that the goal of any scientific community is to reach consensus – broad agreement on which ideas or theories are the correct ones. This is right, but typically consensus formation must first be proceeded by some disagreement about which theories are best. This kind of disagreement, or "transient diversity of beliefs," is necessary to ensure that scientists consider a wide-enough set of possible hypotheses. If they preemptively settle on one hypothesis, without enough exploration, there is no guarantee that they have chosen the best one.[47]

---

[47] See Wu and O'Connor (2023) for an in-depth discussion of this problem in scientific communities specifically.

Of course, too much disagreement for too long is not ideal either. The goal is to explore options for enough time to ensure that optimal theories are selected, without wasting too much time and resources on poor theories. This is related to the explore/exploit trade-off in learning. Note that this "transient diversity" problem is very similar to the "division of scientific labor" problem we discussed in Section 2. The main difference is that we now focus on cases where theories are competing or mutually exclusive. The problem is not how to divide labor across problems generally but how to promote the right level of exploration between competing hypotheses on the same topic.

As mentioned, network models can elucidate different mechanisms that shape diversity of exploration. Note that unlike credit-economy models, the question will not be what *incentives* might impact transient diversity – the agents in these models are typically epistemically motivated – but what other features related to communication and inquiry might do so. The following sections describe different models addressing this question

4.2.1 Limiting Communication

Zollman (2007, 2010) outlines a surprising result. When communities in this sort of network model are less connected, they tend to reach correct consensus more often. In highly connected communities, strings of randomly misleading data can convince every individual to prefer the worse theory. Once this happens, agents no longer test the better theory, and so they do not learn that it is, in fact, better. Less connected communities preserve a transient diversity of beliefs for a longer period of time. In doing so, they spend longer time testing each theory and, thus, increasing their chances of choosing a good one.[48]

This "Zollman effect" is not limited to just these models. It also occurs in network models where actors face NK landscape problems (March, 1991; Lazer and Friedman, 2007; Fang et al., 2010).[49] These will be discussed at length in the next section. They can be understood as search problems where actors look for optima in a landscape that may have multiple peaks. Too much connectivity tends to yield less successful solutions as an entire group adopts the first peak that is found without exploring better ones.[50]

Zollman (2010) argues that this general effect may explain episodes in the history of science such as the premature abandonment of the bacterial theory of

---

[48] For more on this model, see Frey and Šešelja (2018) and Frey and Šešelja (2020) who do extensions and robustness checks.

[49] See also Grim et al. (2015) who find a similar network effect.

[50] Experimental evidence backs up this finding. Real learners in highly connected groups do worse when trying to solve NK landscape-like problems (Mason et al., 2008; Jönsson et al., 2015; Derex et al., 2018).

gastric ulcer disease. One very influential paper finding no evidence of bacteria in human stomachs was enough to turn the entire scientific community away from the correct theory. If, on the other hand, some subpockets of researchers had remained ignorant of this finding, they may have discovered the truth more quickly.

These models imply that decreasing communication in scientific communities might, surprisingly, be a good thing. Zollman (2009) goes further by making one network node a "journal" that publishes results to the rest of the network. He argues that publishing more work is not always better, even when quality is high, because of the Zollman effect. In addition, he argues that journals that randomly publish papers above the threshold may promote transient diversity since randomness will tend to ensure that many different sorts of results are published.[51]

So should scientists communicate less to improve their chances of adopting good theories? Rosenstock et al. (2017) are skeptical. As they point out, decreasing communication only helps when actors face particularly difficult problems, that is, ones where available data do not easily settle the question of what is true.[52] When their problems are easier, there is no need. But scientists often do not know what sorts of problems they face ex ante. As such, decreasing communication might have negative impacts, without the benefits Zollman (2007) identifies. In addition, they point out that there are other mechanisms to promote sufficient exploration in science, some of which will be more efficient (stay tuned).

4.2.2 Stubbornness and Bias

Zollman (2010) points out that if scientists are stubborn, in the sense that they start out with extreme beliefs that take a long time to change with evidence, communities in network models tend to be more successful. This happens, again, because they avoid preemptively settling on a poor theory. Frey and Šešelja (2020) likewise consider agents whose beliefs are sticky, that is, take awhile to change, and find that this stickiness improves group consensus. In a surprising finding, Gabriel and O'Connor (2023) find that confirmation

---

[51] See also Wagner and Herington (2021) who consider restrictions on the sharing of "dual-use research" – work that may have dangerous consequences such as nuclear research or gain of function research in viruses. They argue that such restrictions may not actually hurt scientific progress because they may improve transient diversity of practice by decreasing communication à la Zollman (2007, 2010).

[52] In these models, difficulty corresponds to smaller communities (less data gathered), smaller datasets (less data viewed in each round of simulation), and theories that are more similar in quality (data that tend to be spurious more often).

bias – the tendency to reject evidence that does not fit with one's current belief – can slow learning, increase transient diversity of beliefs, and thus improve group learning.

Mayo-Wilson et al. (2011) use results like these to argue for what they call the "independence thesis" – that individual and group rationality sometimes pull apart in science. At the heart of their argument is the observation that some learning strategies do very well in a group but poorly on their own, and vice versa. One implication is that in thinking about ideal inference methods in science, one needs to be attentive to group-level phenomena. What looks irrational for an individual (stubbornness or confirmation bias) may be rational for a group.

Note, though, that these models appeal to psychological traits that are not necessarily easy to intervene on. In addition, there are harms that might follow from attempts to increase stubbornness or biased reasoning in science (Santana, 2021). For this reason, these mechanisms for transient diversity are also, arguably, poor ones to actively promote.

4.2.3 Grants for Transient Diversity

Another suggestion is that one might coordinate exploration in a community via the choices of grant-giving agencies. If money flows directly to those who explore a variety of hypotheses, centralized bodies can exert some control on the transient diversity of the entire group. Kummerfeld and Zollman (2020) use models and argue that this is a way to get around incentives for scientists to focus on the most promising current theories. Wu and O'Connor (2023) review network models to consider this proposal in-depth and argue that it is one of the most efficient and ethical ways to promote transient diversity of practice in science (see also Goldman [1999] and Viola [2015]). We return to this in Section 7.

4.2.4 Demographic Diversity in Science

Some network models consider how communication structures impact specific subgroups as well as an entire community. Wu (2023b), for example, considers network effects that arise when members of a dominant group systematically ignore or devalue epistemic contributions from members of another group. This kind of "epistemic injustice" could result from racial or gender discrimination, cultural biases, or because economists ignore the other social sciences.[53] She finds that this devaluation has surprising epistemic benefits for the marginalized

---

[53] For more on epistemic injustice and related notions, see Fricker (2007) and Dotson (2011).

group. They tend to learn correct beliefs more often, and faster. This occurs in part because they receive more information but, more crucially, because the dominant group receives less. As a result, the dominant group spends more time testing less promising theories and generating transient diversity of practice that the devalued group can learn from. She argues that listening to the insights of traditionally oppressed groups in science may be valuable for this reason.

Fazelpour and Steel (2022) consider a more symmetric situation where two groups are mutually distrusting. They find, once again, that this kind of distrust can slow learning and, surprisingly, improve community outcomes as a result. They take this result to support demographic diversity in science. The idea is that if there are more subgroups in the community, they will be less likely to preemptively settle on a suboptimal theory. This relates to arguments that diversity can improve deliberation by increasing skepticism and resistance to poor arguments (see, e.g., Sommers [2006]).[54]

This set of results suggests that promoting demographic diversity may improve transient diversity of practice. This particular mechanism for diversity of practice carries few obvious downside risks, is supported by other sorts of arguments, and is arguably an ethical good. For this reason, Wu and O'Connor (2023) argue that along with grant-giving, this is another promising way to promote transient diversity in science.

4.3 Harms of Transient Diversity in Science

As noted, while transient diversity of belief can be beneficial in science, too much diversity of belief is problematic. Eventually scientists should abandon suboptimal theories and adhere to the ones that have proven the most successful. When scientists spend too long on a poor theory, we usually think of this as an epistemic failure. Several network models have illustrated the sorts of things that can go wrong when communities explore theories for too long, and what can lead to these failures.

4.3.1 Scientific Polarization

One result of too much diversity of practice in science is *scientific polarization*, where a community enters a state of persistent disagreement over matters of fact, despite the presence of evidence that might settle the matter. One might naively think that polarization is unlikely to occur in scientific communities. Scientists share evidence and use this evidence to form their beliefs.

---

[54] Fazelpour and Rubin (2022) do complicate these results somewhat by using similar models to argue that homophily can have complex effects on social learning, depending on the types of interactions present.

How, then, would they get stuck in opposing camps? O'Connor and Weatherall (2018) argue that this is actually relatively common and describe a case study. Researchers looking at chronic Lyme disease are strongly polarized and highly mistrusting of those in the other camp, despite apparently aligned interests in learning about Lyme and treating patients.

As noted, stubbornness can beneficially increase transient diversity, but it can also lead to polarization in network models. In both Zollman (2010) and Gabriel and O'Connor (2023), high levels of stubbornness (or confirmation bias) lead to community polarization. When this happens, significant numbers of agents fail to ever adopt good beliefs. Thus, tendencies that are beneficial at the right level can become harmful when taken too far.[55]

Several other network models have investigated the possibility of polarization among scientific agents. Humans are often more likely to trust sources that share their beliefs and identities. O'Connor and Weatherall (2018) consider a network models where agents update more strongly on evidence from those with similar beliefs. In this model, two subgroups can form, each supporting one theory. Since each group only trusts information from their own subgroup, those with false beliefs never learn about the better option.[56] In the group distrust models from Wu (2023b), described earlier, polarization can likewise emerge when one group fully ignores evidence from the other.

Weatherall and O'Connor (2021a) find that conformity can also lead to polarization in network models (see also Fazelpour and Rubin [2022]). Conformist agents prefer to pick the action they think the most promising, but also prefer to match their actions with network neighbors. When this preference is strong enough, cliques of actors can form, who each prefer different theories. In addition, a number of network models find that conformity in general harms social learning because it leads actors to ignore good information and fail to pass on their knowledge to others in a group (Mohseni and Williams, 2021; Weatherall and O'Connor, 2021a; Fazelpour and Rubin, 2022; Fazelpour and Steel, 2022).

This literature suggests that transient diversity can go too far. Another takeaway is that there are multiple mechanisms whereby scientific polarization might arise, all of which are at least somewhat psychologically realistic. If scientists do not trust those with different beliefs, if they (like the rest of us) wish

---

[55] This observation lends credence to claims that stubbornness is a poor way to promote transient diversity of practice.

[56] Weatherall and O'Connor (2021b) extend this work to consider how multiple beliefs might polarize to create epistemic factions, but this Element mostly focuses general beliefs rather than scientific communities. Olsson (2013) models Bayesian agents who gather data and can polarize as a result of mistrust, though his model focuses on testimony rather than evidence sharing.

to conform with peers, if they engage in confirmation bias, if they are irrationally stubborn, or if they do some combination of these things, they, too, might fall into polarized camps. And when this happens, some scientists persistently fail to develop true beliefs.[57]

4.3.2 Industry, Policymakers, and Networks

To this point, we have considered network models where all the actors have epistemic motives. They wish to discover truth and use evidence to do so. In many cases, though, scientists also communicate with agents who hold less pure motives. Sometimes, for example, agents with industrial or political motives will engage in epistemically detrimental dissent, or dissent purely aimed at confounding public belief (Oreskes and Conway, 2011; de Melo-Martín and Intemann, 2018). In doing so, they can create diversity of practice that is strictly harmful.

Holman and Bruner (2015) consider what happens when an industry agent tries to sway consensus in a scientific community. In their model, one agent persistently shares data that push toward an inaccurate belief. This could represent, for example, a tobacco-funded scientist sharing fraudulent studies that find no link between tobacco and cancer. They show that in this sort of case, networks of trusting scientists can be prevented from reaching accurate consensus by these misleading results.[58]

Weatherall et al. (2020) consider interactions between science, industry, and policymakers. In their model, an industry actor purposefully cherry-picks evidence from the scientific community to share with policymakers. As they show, even real, unbiased evidence can be used to confound policymakers. They explore the conditions that make these processes easier or harder. Whenever studies are more likely to be wrong, there is more fodder for propagandists. They argue for high-quality standards in science, such as requiring large sample sizes, on the basis of this reasoning.

Lewandowsky et al. (2019) develop a similar, but in some ways more realistic, model of climate change denialism. They consider three interconnected networks of (1) scientists, (2) industry denialists, and (3) the public. Agents observe real global temperature data and update their beliefs about whether the globe is warming on the basis of it. Denialists in their model can share short,

---

[57] For a very different modeling approach to polarization, or "fragmentation" in science, see Balietti et al. (2015) who use an epistemic landscape model and agents who weigh influences of both truth and social forces.

[58] Though if they add trust dynamics so that scientists can learn to ignore the industry agent, the community does better.

unrepresentative, cooling trends with the public in support of the false position, thus confounding public belief. Together these three models suggest that attempts to control information sharing by motivated agents may be effective. In doing so, they support qualitative claims about the effectiveness of industry strategies to promote detrimental dissent.

Wu (2023a) uses network models to ask another question about industry research: Can industrial scientists gain an advantage by failing to adhere to the communist norm? While academic scientists share their work freely, most industrial research is proprietary. In her models, one group is able to receive all evidence produced in the network but does not share their own evidence. She finds that this group learns more quickly and accurately.[59] If academic researchers spend more time exploring possibly suboptimal options, industrial researchers can learn from their exploration while also exploiting current knowledge to their own advantage. This is an under-explored possibility for how industry might exploit science to its benefit. On the basis of this research, Wu and O'Connor (2023) suggest that requirements should be implemented for industry sharing.

4.4 Scientific Network Formation

We have seen how scientific network structures crucially impact group learning. A natural question is: How do scientific networks form? And given the principles of scientific network formation, what sorts of social structures naturally occur in scientific communities?

4.4.1 Preferential Attachment

Barabâsi et al. (2002), in an influential paper, attempt to answer this question for co-authorship networks where each link represents a collaborative partnership between scientists. They start with empirical data on collaboration networks, which reveals certain characteristic structures. First, like most human networks, collaboration networks are "small-worlds," that is, have cliques and short average path lengths between individuals. Second, collaboration networks follow something like a power-law, meaning that a small number of scientists have large numbers of collaborators, while the majority have just a few collaborators.[60] Barabasi et al. argue that this pattern emerges as a result of *preferential attachment*, meaning that new individuals in a discipline are more likely to collaborate with scientists who already have more collaborative links. They use

---

59 The mechanism is similar to the one that benefits marginalized groups in Wu (2023b).
60 See also Newman (2001, 2004); Glänzel and Schubert (2004).

their results to develop a preferential attachment model of the formation of collaboration networks and show how it can replicate the real-world data.[61,62]

Anderson (2016) looks at an explanation for why scientific collaboration networks might show preferential attachment patterns. She is interested in how the particular skills and properties of individual scientists might impact collaboration choices. In her models, scientists have different skill sets and seek others with complementary skills for collaboration. She shows how differences in skill can lead to surprisingly stark differences in connectivity, where rare individuals with useful skill combinations have large numbers of collaborative partners. On this picture, preferential attachment in science actually plays a functional role. (Though alternative explanations might appeal to the personalities of some coauthors, or their popularity, meaning preferential attachment might not arise for any obviously epistemic reason.)

4.4.2 Homophily

Social networks generally are sensitive to social identity. For instance, most social networks are homophilous in various ways – meaning that individuals prefer friends, contacts, and partners who are like themselves. How does identity impact networks of collaboration and communication in science? And what impacts does this have on fairness, credit-giving, and scientific progress?

Rubin and O'Connor (2018) use network models to consider whether such patterns might arise as a result of discrimination in collaborative bargaining.[63] If women or academics of color receive less credit per work performed when collaborating with men or white academics, they may learn to avoid such collaborations. In their models, they show that if unfair patterns of collaboration emerge, those who are disadvantaged tend to avoid discriminators as a result.

---

[61] This does not exhaust their findings, but, for space reasons, I do not describe their other results. More recently, a number of authors have generated insights into scientific networks by analyzing data on coauthorships. We do not go into depth here, but see, for example, Sampaio et al. (2016) and Bender et al. (2015).

[62] Jackson and Wolinsky (2003) also give an early model of the process of coauthorship network formation. They show that if authors work more synergistically when they have fewer coauthors, academics may sometimes choose an inefficient surfeit of collaborative partners. Their main goal is to show how sometimes stable networks will be inefficient with respect to agent payoff.

[63] In several disciplines, it has been found that women are less likely to collaborate than men and are more likely to collaborate with other women if they do so (Ferber and Teiman, 1980; McDowell and Smith, 1992; Boschini and Sjögren, 2007). And although there is less research into whether similar patterns hold with respect to race, Del Carmen and Bing (2000) find that black criminologists are less likely to coauthor than whites, and Botts et al. (2014) find that black philosophers strongly cluster in subdisciplines.

This leads to homophilous networks where marginalized academics collaborate with those in their in-groups more often.

Schneider et al. (2022) consider how to reverse this trend. They argue that in some cases, there are benefits to diversity in collaborative groups and that these benefits may be lost in homophilous scientific networks. As they point out, special incentives may promote diverse collaborations but, in doing so, may lead to instances of discrimination by bringing diverse groups into contact.

• • •

In thinking about how to shape scientific communities, communication structure is a key feature, as we have seen. In particular, models help illuminate large-scale, emergent phenomena that arise in networks as the result of patterns of communication, especially those surrounding diversity of practice. The next section will keep this focus on emergent phenomena in scientific groups, while further considering how scientists select research problems.

5 Epistemic Landscapes

Nicolas has just started his first faculty job. He is considering what sorts of research areas to move into as he wraps up his dissertation work. In general, he is someone who does not like going along with trends. His friends describe him as an outside-the-box thinker. He ends up developing a research project on self-mutilation related to relationship stress: A topic that some of his colleagues find strange or distasteful but is certainly something that few others have yet bothered to investigate.

Liam is likewise setting up a research program at his first faculty job. While creative and meticulous, he is less inclined toward risky research projects than Nicolas. He decides to extend the work of his graduate advisor, who studies political patterns related to Marxism. He starts a project on the impacts of Marxist thought on science funding.

• • •

Before scientists run experiments, build models, communicate their work, or draw inferences, they must decide what topics to work on. This decision process involves many factors – their expertise, the distribution of work currently done in the community, available funding, availability of various resources (FMRI machines, experimental subject pools, expensive telescopes, hadron colliders, etc.), and also the individual interests and personalities of the scientists involved.

We have already seen a number of models that look at how scientists choose topics for research. Credit-economy models of the division of labor consider how priority incentives push scientists to choose topics of study based on their promise and the distribution of work in the community. Selective models consider how self-preferential biases might promote the study of currently popular topics or paradigms. Network models show how communication structures might change scientists's beliefs about various approaches and, thus, shape their research choices.

In this section, we consider another model of problem choice in science – the epistemic landscape model. In this sort of model, scientists move about on a landscape where locations represent different problems or approaches in science that may have more or less epistemic merit. As we will see, it is a paradigm that allows us to consider the way individual personalities and interpersonal traits may shape research choices. To this point, the modeling paradigms we have considered treat scientists as a uniform group – they maximize credit expectations similarly, respond to selective pressures similarly, or communicate similarly. Epistemic landscape models, on the other hand, focus much more on the relevance of individual cognitive style to problem choice. In addition, as will become clear, compared to the models discussed so far, epistemic landscape models are more focused on how scientists might explore new, untested research possibilities (rather than dividing their time among current ones).

5.1 Epistemic Landscapes and Exploration

As discussed throughout this Element, diversity of various sorts can impact the functioning of scientific communities. Weisberg and Muldoon (2009) introduced an epistemic landscape model to think about how cognitive diversity, that is, diversity in the preferences and personalities of scientists, might impact division of labor.[64] Their model involves a two-dimensional landscape with several peaks, including one global maximum. Figure 2 shows an example of what such a landscape might look like.[65] Different locations on this landscape are taken to represent "approaches" in science. Weisberg and Muldoon think of approaches as including a research question, instruments and techniques used in data gathering, methods of data analysis, and background theories used to interpret data (228). Similar locations in the space represent similar approaches. They point out that given the complexities of real-world research approaches, more realistic landscapes would be highly multidimensional, though they focus on just two for simplicity reasons.

---

[64] Their model has some similarities to fitness landscapes as first developed in biology (Wright et al., 1932) and to foraging models.

[65] This figure does not specifically represent the landscapes used in Weisberg and Muldoon (2009), who focus on two-peaked landscapes with all other locations entirely flat. But other models of epistemic landscapes take more general approaches.

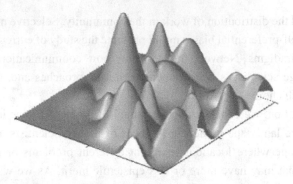

Figure 2 Example of an epistemic landscape

The height of the landscape in each location represents something like the epistemic quality of the approach – the ability of that approach to contribute to knowledge generation. Weisberg and Muldoon assume that changes in significance are smooth, that is, that similar approaches tend to yield similar levels of epistemic success. Agents in the model represent scientists searching this space for insights. An agent located on a patch is currently employing that approach. Search practices can vary as a function of personal characteristics, features of the landscape, or in response to what others in the community are doing.

As noted, Weisberg and Muldoon develop the model to investigate how cognitive diversity among academics – different preferences, backgrounds, styles, and so on – might promote division of labor. Note that here division of labor is not just a problem of choosing between existing, well-understood problems to solve (as with credit-economy models) but also a problem of searching for new, important topics of research. They argue that a combination of different sorts of academics – "mavericks" who like to work on new, unusual approaches and "followers" who like to repeat the approaches of others – search the landscape better than a uniform group. Their work has been convincingly critiqued as failing to show this result (Alexander et al., 2015; Thoma, 2015; Pöyhönen, 2017; Pinto and Pinto, 2018). In actuality, uniform groups of mavericks do best in their model since they manage to explore the space more successfully.[66] But their research set the stage for further models that do support this finding.

Thoma (2015) shows just this. She looks at epistemic landscape models where academics can use "explorer" and "extractor" type search strategies (similar to mavericks and followers). Her explorers prefer to do work that is very different from other agents – like Nicolas. Her extractors prefer work that is similar to other agents but still novel – like Liam. Both sorts of scientists

---

[66] Relatedly the follower agents in their models were unrealistically poor explorers.

also prefer to do research that is of higher significance and incorporate this preference into their search. In addition, she assumes that "movement" on her landscapes is semilocal.[67] Scientists are constrained by their training, equipment, and knowledge from moving to patches that are too distant. She finds that communities do best to have some combination of both explorers and extractors.[68] Explorers chart new areas of significance, while extractors follow them to these areas and research the entire space.

Pöyhönen (2017) further supports this claim in a variant of the epistemic landscape model where the significance of an approach is "depleted" as agents visit it. The idea is that each approach will yield some number of important findings, and the goal is to have various agents employ that approach long enough to produce them. He considers a different sort of cognitive diversity, represented by different rules for social learning. Agents prefer to hill-climb on the landscape but will potentially reorient if they observe peers who take better approaches. Agents differ in how similar the approach of the peer must be, and how much better, before they will mimic them. He finds that on landscapes with many "dead ends," that is, small, local maxima, a diversity of learning strategies does best. Individualists explore their own areas of the landscape without all herding onto the same problems. But social learning allows agents to eventually escape local maxima and find areas of higher significance.

In the last section, we saw how disconnected groups can perform well in science by dint of preserving transient diversity of practice. Grim et al. (2013) consider an epistemic landscape model where the problem space can vary in difficulty. Some spaces are smooth, with just a few (or even just one) peak, while others are more rugged and multipeaked. In the latter cases, search is more difficult. They consider the question of how communication structures impact search by supposing that scientists are situated both in the landscape and an epistemic network. Agents can then move across the landscape by copying those they communicate with. Reflecting the results on diversity of practice described in Section 4, they find that less connected groups are better at finding global peaks.

Adding robustness to all these findings, Devezer et al. (2019) consider a group of scientists who search a space of possible models for the best fit to data they gather. Groups of scientists employing different search strategies tend to outperform uniform groups (though exploratory scientists who

---

[67] As opposed to Weisberg and Muldoon (2009) whose agents can only move to neighboring patches.

[68] This finding assumes that scientists have at least some flexibility in what approach they choose, that is, that their movement in the landscape is not too constrained.

try out very different models from the current one seem to be most important to group success, reflecting the importance of mavericks in Weisberg and Muldoon [2009]).

In critique of these findings supporting cognitive diversity in science, Alexander et al. (2015) point out that there are search strategies that are uniform but do very well in epistemic landscapes. One of these is a "swarm" where flocks of agents respond both to the height of the landscape and to the success of nearby agents. The suggestion is that an especially good search strategy can find epistemic significance without needing different sorts of agents. But it may be that in real epistemic communities, "swarm" strategies are not particularly realistic to employ, and, thus, exploring the interactions of different scientific personalities in problem choice may still be worthwhile.

5.1.1 Lotteries

Since epistemic landscape models allow us to represent exploration of a wide problem space in science, we might use them to ask: What funding strategies are successful in science? And, in particular, what funding strategies promote the right amount of exploration in an epistemic landscape?

Avin (2015, 2019) uses epistemic landscape models to address these questions. As discussed in Section 4, which topics are explored in a scientific community is typically decided by the interests of individual scientists and also by the decisions of funding bodies. In his model, agents receive funding from a central agency to research some landscape patch for some length of time. Agents without funding enter a pool where they do not explore the landscape but instead seek a new grant to research the highest patch in their local vicinity. In addition, new researchers seek funding to explore new, randomly chosen research areas. Funders are able to estimate the epistemic quality of approaches that are near those already explored (and thus where there is information about similar projects) but not new areas. Across a variety of different assumptions, Avin finds that some randomness in the funding process significantly improves community exploration. In particular, a funding strategy that chooses some of the most promising projects, and some random projects, does very well. This is because random funding promotes beneficial exploration, much like the presence of exploratory types of individuals does. Funding can thus replicate the impacts of exploration and exploitation strategies used by individual scientists. On the basis of this research, Avin advocates lottery funding in science.

Harnagel (2019) builds a more realistic model to test this idea. She uses citation data to build her epistemic landscape. Subdisciplines (such as "crop science") are nodes located within a network that tracks similarity of references.

Height in this network tracks citation count. Her agents do not move in this landscape, but otherwise her dynamics are similar to those of Avin – new researchers and those without funding apply for grants and perform research at their location. Her results, like Avin's, suggest that randomness introduced by lottery funding improves community output.

The suggestion in Section 4 was that grant funding might be used to deliberately promote diversity of practice in science. The suggestion here is that *randomness* in grant funding could do the same. Note, again, though, that the focus here is more on how scientists find new topics for exploration, rather than how they should split time between different competing theories.

5.1.2 Interpreting Epistemic Landscapes

Although I have not gone into much detail, these different uses of the epistemic landscape model disagree about how best to interpret it. Should each approach represent a fine-grained problem that only one scientist can work on (Thoma, 2015) or a more general topic that multiple scientists can tackle (Weisberg and Muldoon, 2009; Pöyhönen, 2017)? Is the goal for a community to discover the area of peak significance (Weisberg and Muldoon, 2009; Grim et al., 2013), to discover all significant areas (Weisberg and Muldoon, 2009; Thoma, 2015), or to "mine" different areas for findings based on their significance levels (Pöyhönen, 2017; Avin, 2019)? Should scientists be able to move only locally (Weisberg and Muldoon, 2009; Avin, 2019), semilocally (Thoma, 2015), or with varying levels of flexibility (Pöyhönen, 2017) or widely across the landscape (Grim et al., 2013)? Does the landscape maintain its shape (Weisberg and Muldoon, 2009) or change as it is researched (Pöyhönen, 2017; Avin, 2019)?

One worry about this sort of model is that this disagreement reflects a lack of clarity about exactly what is being modeled and how. While models are often employed flexibly across different arenas, in this case, the models are being applied to the same arena but with disagreement about what the elements correspond to. For this reason, it makes sense to treat findings from this literature with particular thoughtfulness, paying careful attention to the structures and interpretations in the particular model.

Another concern, raised by Alexander et al. (2015) and Bedessem (2019), is that the two-dimensional landscape approach fails to capture the complexities and interdependencies of most scientific theory spaces.[69] For example, sometimes multiple aspects of an approach might be synergistic in surprising ways

---

[69] This is also a worry for Harnagel (2019) who, as noted, addresses it by using empirical data to shape her landscape model.

so that a small movement in one dimension might cause big changes in the overall quality of the approach. Sometimes approaches have surprising benefits to future research that are not initially obvious. For these reasons, more complex landscapes might provide better models of some scientific disciplines.[70]

5.2 Rugged Epistemic Landscapes

How might we go about representing scientific theory space with a more complex landscape? Alexander et al. (2015) argue that, in particular, two things are missing from epistemic landscape models, (1) the fact that it is often hard to leave "local optima" in science, and (2) there are often complex interdependencies between aspects of an approach (as noted earlier). For this reason, they suggest that modelers should instead represent science with NK landscape models.[71]

As noted in Section 4, NK landscapes are similar to two-dimensional ones in that they have locations that are each associated with some value, where the goal is generally to find higher value areas. Their key features are that they are highly multidimensional and "tunably rugged."[72] By varying parameters, one can shape the amount of interdependency between different dimensions of the landscape and create either smooth or rugged features. Figure 3 shows three schemata meant to demonstrate the difference between more smooth and more rugged NK landscapes.[73] Importantly, rugged landscapes have many local optima that are hard to escape. We can think of this as corresponding to problems where local success can lead searchers down a path that then tends to cut off access to even better approaches.

---

[70] Another worry, raised by Pinto and Pinto (2018), regards robustness of results in these models. They point out that small changes to the exploration algorithms significantly alter the success of different agents in the model. This raises the question of whether these results should be taken to bear on real-world communities.

[71] These were first developed by Kauffman and Levin (1987) and Kauffman and Weinberger (1989).

[72] A location in these landscapes is determined by the values of a binary string of length N. For instance, the location of an agent might be <0,1,0,0,1,1,1>. Thus, the landscape is N-dimensional, with two possible locations along each dimension. We can understand these entries, *pace* Alexander et al. (2015), as tracking binary beliefs in a set of scientific propositions or the use of scientific methodologies. The first entry might track whether or not a scientist uses a 0.05 significance threshold, the second whether they p-hack, the third whether they think smoking causes cancer, and so on. Each of these contributes to the overall fitness of the approach, but the contribution of each can depend on the values of the other entries (for instance, p-hacking might be especially bad for those who also use the 0.05 significance threshold). The value of K determines the level of interdependence, such that if K is higher, the fitness of one aspect of an approach depends on a greater number of other aspects.

[73] This figure should be understood as an aid to understanding rather than a literal representation. I draw from Lazer and Friedman (2007) in developing it.

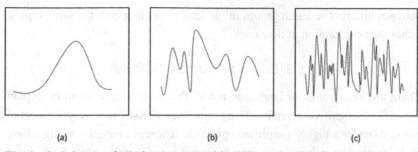

(a)   (b)   (c)

Figure 3 Schema of NK landscapes with different levels of ruggedness: (a) a smooth landscape; (b) a more rugged one with multiple peaks; (c) a highly rugged landscape with little correlation between location and height

Alexander et al. (2015) use these models to argue that the entire research program described earlier – in which social influence between agents is an important part of problem choice in science – depends on the assumption that the epistemic landscape is somewhat smooth. If so, agents can find areas of significance by observing peers. Extractors can follow exploiters to new, interesting topics. If the landscape is too rugged, though, this social learning will not be helpful because a peer's success does not lead an agent to worthwhile, nearby topics. In most scientific disciplines, though, the epistemic landscape is arguably "smooth" enough that social information about problem choice is relevant and helpful.

NK landscape models have not only been used for skeptical projects, though. Some work on these landscapes further supports the main claim above – that diversity is crucial for scientific exploration. In Section 4, we discussed the work of Lazer and Friedman (2007) and Fang et al. (2010). They implement models where networked agents explore NK landscapes and find that less connected groups tend to be more successful. Likewise Boroomand and Smaldino (2021) and Wu (forthcoming) find that randomness in search strategies in NK landscapes can improve group outcomes, by allowing actors to escape local optima.[74] All these models look at mechanisms that introduce variety into the exploratory strategies agents employ and show that this variety helps exploration. Note, however, that these papers interpret NK landscapes in a different way. The landscape represents a single, complex problem that a group attempts to solve (rather than locations representing different problem choices in science). For this reason, their results are arguably most germane to thinking about transient diversity across competing theories in the same domain. However, they could also be reinterpreted to suggest that too much social influence

---

[74] See also Barkoczi and Galesic (2016) and Yahosseini and Moussaïd (2020).

between different research groups might lead to herding onto the same topics, rather than exploration of new ones.[75]

5.3 Diversity Trumps Ability

There are other forms of landscape model that have been used to champion the case of cognitive diversity. Hong and Page (2004), in a very influential work, consider a highly simplified epistemic landscape model – a ring where each location on it is associated with a random epistemic value. Each agent consists of a list of small integers, representing their search "heuristics." An individual with the list <4,7,12>, for example, is able to view locations 4, 7, and 12 steps in front of their current place. They travel around the ring, using these heuristics, and moving forward whenever they can access a position of better quality. Agents stop when no visible option is an improvement on their current location.

After randomly generating a set of agents, Hong and Page test their average success across all possible starting positions. This yields a success metric for individuals and generates a group of "experts" – those individuals that do best on average. Their central finding involves comparing the success of groups of experts versus more "diverse" groups. (A group solves the problem by using their heuristics in sequence until they cannot improve their position.) They find that, on average, diverse groups do better because experts tended to have a lot of overlap in their heuristics. (For example, 3 might be an important heuristic on a particular landscape, and all experts might have 3.) This means that expert groups tend to have fewer ways to explore the space than diverse groups. Their findings have been used widely to argue for the benefits of diversity, and even for the importance of democracy (Landemore, 2012).

There have been further developments of this model and debates about its applicability.[76] One general concern is that this model and its search rules are even more simple than those discussed so far. For this reason, it arguably does not have enough structure to represent the relevant factors in determining how diversity is important to group learning.[77] We do not overview this literature in depth here. However, it is worth noting that these models are usually taken to

---

[75] Heesen et al. (2019) provide a very different model supporting "methodological triangulation," or the use of different approaches to one problem. In their model, each method can be thought of as a partially independent vote in favor of a finding. They appeal to the Condorcet jury theorem to argue that more votes should generally imply better confirmation of a theory.

[76] See, for example, Thompson (2014) and Singer (2019) over whether randomness or diversity explain group success in the model. See Reijula and Kuorikoski (2021) for critiques of the framework.

[77] For example, Grim et al. (2019) argue that its structure does not allow for a reasonable definition of "expertise" among agents.

further support the idea that cognitive diversity, that is, diversity in approach to problems, is important in epistemic groups.

• • •

To this point in the Element, we have seen many variations on a theme regarding the promotion of diversity of practice in science. Landscape models address the best ways to explore new research topics, and how to divide labor between this sort of exploration and the in-depth work of exploring known approaches. The mechanisms considered mostly appeal to cognitive diversity (but also to funding rules, and rules for social learning). While the general take-away from this group of models about the importance of cognitive diversity is likely correct, we have seen that there are some concerns about the representational adequacy of these models to support this claim. In the next section, we take a pivot away from a focus on problem choice to look at the replication crisis, methodological practice, and interventions intended to improve inquiry.

6 The Replication Crisis and Methodological Reform

As a graduate student, Firenze learned from his advisor that when his original hypothesis was not supported by the data, the next move was to analyze the data for alternative findings. He would look at each demographic factor in his study population to see whether his hypothesis would hold in any subgroup. He would hunt for alternative explanations for his data. After inevitably finding something, Firenze would then write up a paper implying that the original hypothesis actually matched the post hoc finding. Firenze did not see anything wrong with this – after all, his findings were, in fact, supported by his datasets and he was not committing fraud. Upon becoming a PI himself, Firenze trained his students to do the same.

Isobel was one of several PIs searching for a link between genetics and sexual orientation. Her team considered thousands of possible links along these lines in a large dataset. She found a number of statistically significant associations and reported them in a published paper. Like Firenze, she treated her findings as if they were ex-ante hypotheses, rather than ex-post connections combed from a large dataset. The press had a field day, widely sharing her findings.

Yolanda gathered a dataset on whether infants had longer gaze times for attractive faces than unattractive ones. Her initial analysis did not yield a finding; so she decided to run some more subjects. She checked for significance after every five babies she tested. After getting a significant finding, she stopped testing and published her results.

• • •

It is hard to overstate the impact of the replication crisis on the sciences. This "crisis" began with a number of highly publicized failures to replicate a large number of prominent findings, especially in psychology and biomedical sciences (Begley and Ellis, 2012; Open Science et al., 2015; Baker, 2016; Eklund et al., 2016; Camerer et al., 2018). It has upended many disciplines' self-conceptions, overturned common methodological practices, led to widespread reforms, and initiated a huge turn toward the study of metascience.

In line with this movement, many authors have used models to ask what sorts of practices are most harmful to inference, and what practices would be better. Many of the questionable practices that have drawn attention are similar to those that Firenze, Isobel, and Yolanda engaged in – not outright fraud but data fiddling that illegitimately increases the chances of yielding a positive, publishable finding.[78]

This section of the Element looks at literature mostly developed in response to the replication crisis on questionable practices, better practices, and optimality in inference from data. Here I deviate from the rest of the Element by organizing around a topic, rather than a modeling paradigm. For this reason, I include more work that is theoretical even if it does not present novel modeling results.[79] In addition, the models in this section of the Element tend to focus less on emergent group aspects of science and more on how inferential practices impact individual findings. But, as we will see, many features of community incentive structures play key roles in these models of individual research practices and proposals to improve practice are mostly focused on changes in communal norms and rules. As in the rest of the Element, the models correctly assume that human features crucially shape the production of scientific knowledge.

6.1 Are Most Findings False?

Ioannidis (2005), drawing on previous literature, presents a highly influential model of scientific discovery and argues on the basis that most research findings in scientific fields are false. He starts by analyzing the probability that a claim in a scientific literature reflects a true fact, given current statistical practices (and

---

[78] The downfall of nutrition researcher Brian Wansink provides a concrete real-world example. Questions about his research practices first started emerging after his blog posts that described p-hacking and HARKing (Dahlberg, 2018).

[79] Much of this work employs statistical formalisms or frameworks, though. In a few rare works of metametascience, Devezer et al. (2021) make the argument that metascientists should be generally employing formal methods in their arguments in order to ensure rigor, and Gorman et al. (2019) similarly suggest that investigating interventions in science should take a "systems approach." These papers stand in support of the work reviewed here and throughout this Element.

especially NHST. More on this shortly). This depends on a few factors. First, it depends on the prior probability of the claim's truth, that is, its likelihood with no information about study findings. Ioannidis points out that various fields will have different probabilities, or base rates, of true hypotheses depending on a host of factors including how they generate hypotheses, and the sort of topics they work on. Second, the likelihood that a finding is true depends on the statistical power of the study, that is, the ability of the test to correctly identify a true claim as true. And, third, the likelihood also depends on the level of the statistical significance threshold (i.e., the propensity of a test to misidentify a false claim as significant). Ioannidis builds a very simple model assessing the probabilities that scientific tests will correctly identify true or false facts, given these three factors.

He identifies a number of implications of this simple model. (1) Small studies, which have lower statistical power on average, are more likely to label a false claim as significant (a point that has also been made extensively elsewhere). (2) Fields with small effect sizes will have more false findings because it is more difficult, in these fields, to disambiguate true and false claims. (3) Fields with more room for bias, that is, with fewer constraints on standard research practices, will have more false findings. In these cases, researchers incentivized to publish have more flexibility to use research practices that allow them to find positive associations, whether or not such associations are real. (4) Fields with many low-probability hypotheses, rather than fewer high-probability ones, will likewise have more false findings. This role of prior probability of hypotheses in study outcomes has been particularly underappreciated. Many scientists think hypothesis generation can be highly flexible, as long as the hypotheses face rigorous tests. Ioannidis's model forcibly refutes this idea.[80] (5) Ioannidis further considers the possibility that multiple teams test the same hypothesis. If at least one yields a positive finding, it will be reported in the literature, increasing the chances that any reported finding will be false. Assuming strong publication bias, the more focus there is on one question in the literature, the more likely *some* team is to yield a positive finding regardless of truth.

On the basis of this analysis, Ioannidis considers features of a number of disciplines and concludes that, surprisingly, their likelihoods of a reported finding being true are often very low, sometimes less than 1 percent. This is especially noteworthy in fields with low base rates of true hypotheses, such as exploratory biomedical research.

---

[80] For more on the importance of base rates of true hypotheses, see Smaldino (2019b) and Bird (2021).

Subsequent authors have argued that some of Ioannidis's assumptions are too strong, thus painting an overly negative picture of scientific research (Goodman and Greenland, 2007). The general message, however, holds. Even though no form of statistical inference will be correct in every case, most scientists trust statistical practices to properly constrain findings. But as Ioannidis shows, the structure of our primary method of inference does not always work the way we have traditionally assumed it does.

In a complimentary work, Ioannidis (2008) argues that most true associations discovered in the sciences are inflated, that is, that their effect sizes are overstated. He uses a simple simulation to show that this will happen when (1) scientists must meet a significance threshold to report results and (2) studies are generally underpowered. In these conditions, those tests that reach significance will also tend to have larger effect sizes, leading to a general overestimation in the literature. He shows that the situation is exacerbated when researchers have flexibility in their choice of analysis that can impact effect size. In such cases, the studies with the largest estimated effect sizes tend to be the ones that are published and shared, potentially leading scientists to strongly overestimate effect sizes.

6.2 Self-correction and Publication Bias

Savvy consumers of scientific data typically do not assume that one positive finding establishes the truth of a claim.[81] In many cases, it is warranted to strongly believe a scientific claim only after it has been appropriately replicated and meta-analysis performed on the results. (This, remember, is why we saw various proposals for credit incentives to increase replication in Section 2.) This approach to inference assumes that the sciences are self-correcting. While one study may be erroneously misleading or while a literature may temporarily go down a wrong track, eventually as more data is gathered, the community corrects toward better understanding.

Moonesinghe et al. (2007) develop a model based off the one from Ioannidis (2005), but where a study is performed some number of times, n, and r tracks the number of these that reach significance. As they point out, the higher the r is, the greater the chances that the claim is true. They conclude that a little replicability goes a long way toward establishing the truth of a scientific claim. They note that successful replications are only likely to work when studies are of high enough quality, that is, the power (ability to accurately detect a true result) is high enough, and when biased production is not a significant problem.

---

[81] There are some exceptions. Some types of scientific evidence provide conclusive, or else extremely strong, data in favor of a conclusion. Usually, data are more equivocal.

Their model assumes that it is possible to observe all successes and failures of some scientific test. In this way, one can correctly calculate both r and n and get some ideas about the likelihood that a claim is true. But because of publication bias toward positive, novel findings, it is rarely possible to assess how many tests of a hypothesis have been performed. This is sometimes referred to as the "file drawer" effect – scientists run studies, and if they obtain null results, the data go into the file drawer (Rosenthal, 1979). This is both because journals do not tend to accept null results and because scientists, anticipating this, do not tend to submit them for review. Along these lines, Fanelli (2012) estimate that more than 80 percent of published papers report positive findings.

Building from the models just discussed, Nissen et al. (2016) consider a process whereby scientists test and retest a claim multiple times to establish its accuracy. They assume members of a community engage in Bayesian inference to draw conclusions based on published results, and they ask when and how false claims can become "canonized" or understood to be so likely that they are no longer tested. Similarly, Romero (2016) presents a simple simulation where scientists repeatedly replicate a study. Both papers consider how publication bias will impact beliefs in this sort of scenario. Without publication bias, and with enough repetition of tests, Romero's scientific community will always "self-correct," that is, eventually develop an accurate picture of the size of some scientific effect. The community in the Nissen et al. (2016) model eventually stops testing claims that seem highly likely or unlikely, and so they are not guaranteed to reach accurate beliefs but typically do so without publication bias. Both models, though, find that publication bias seriously harms the ability of a community to develop accurate beliefs. In Romero's model, publication bias leads to persistent overestimation of effect sizes. In the Nissen et al. (2016) model, they find that across a range of conditions, false beliefs are canonized as a fact unless null results are regularly published. Thus, these two models show how even replication work may fail to identify false associations when there is publication bias. (Though Bruner and Holman [2019] point out that in Romero's model, using statistical tools to identify and account for publication bias can allow for self-correction in the face of publication bias.)

Even more pessimistically, Devezer et al. (2019) develop a model where groups of scientists compare possible models of the world. They reject those models that are worse fits to data they gather. Some scientists replicate recent comparisons, while others try new ones. These authors find that even without publication bias, replication does not ensure that the community comes to correct consensus on the best model. Instead, successful replications of tests of a decent, if not ideal, model can mean the community spends a lot of time

focused on a suboptimal theory. Altogether, these results lend further weight
to the worries raised by Ioannidis.

6.3 Questionable Research Practices

One focus of the model from Ioannidis (2005), and follow-ups, is researcher
flexibility. A key worry about this sort of flexibility is that it allows researchers
to use QRPs. These are practices that make it easier for researchers to publish
positive findings, even in cases where these findings do not reflect the under-
lying reality. One core part of the response to the replication crisis has been to
identify these practices and intervene to prevent them. The literature on these
interventions includes both modeling work and technically informed argumentation. This section will discuss both sorts of work, though for space reasons it
will be a very selective review.

　　Many of the most worrying QRPs involve some sort of flexibility in how
researchers manipulate data and statistical tests. Via these manipulations, they
increase their chances of obtaining significant positive findings (Simmons
et al., 2016). These manipulations are not necessarily fraudulent in the sense
that researchers are employing them purposefully to benefit their careers at
the cost of research integrity. In many cases, researchers report having learned
to employ QRPs from their advisors and peers (as Firenze did), without a
sense that the techniques they were learning might be problematic. Before discussing possible responses to QRPs, let us take a little more space to understand
them.

6.3.1 P-Hacking

P-hacking encompasses a set of practices whereby scientists perform a large
number of statistical tests on data and then report just the ones that are significant (Selvin and Stuart, 1966; Smith and Ebrahim, 2002; Head et al., 2015).
Suppose a researcher hypothesizes that vaccines cause autism. Suppose also
that, like Firenze, their data do not end up supporting this hypothesis. One
option is to give up on the study or try (perhaps unsuccessfully) to publish a
null result. But these options are not good ones for a researcher seeking credit.
Instead they might look for an association in the data among just female or
just male children. If they do not find one, they could instead look for an association among just female children with wealthy parents, just male children
with wealthy parents, just female children with poorer parents, and just male
children with poorer parents. Suppose they do find a vaccine–autism association among high socio-economic status females, and report this as the positive
finding of their work.

This is just one example of how p-hacking can work, but it illustrates how performing enough tests can almost ensure that a researcher finds a significant result, though this result is likely to be a false positive. The umbrella of p-hacking encompasses multiple specific practices (Head et al., 2015) including, but not limited to, (1) analyzing data for significance throughout a study and selectively stopping when significance is reached, as Yolanda did (John et al., 2012), (2) tweaking data or "fiddling" when a finding is marginally significant, that is, by dropping outliers (Gadbury and Allison, 2012), (3) deciding after statistical analysis to combine or split groups of data, and (4) doing this last until a significant result is found (as in the example above) or "fishing."[82]

Relatedly, Gelman and Loken (2013) use a simple model to describe what they call "the garden of forking paths." This involves researchers altering their tests based on the type of data they observe in ways that increase chances of obtaining significance. This need not involve performing multiple statistical tests as in p-hacking, but it rather involves selecting the most perspicuous statistical test for the sort of data gathered, thus increasing the chances of yielding positive results.

6.3.2 HARKing

Another widely discussed QRP is HARKing, or hypothesizing after results are known (Kerr, 1998). This includes a cluster of behaviors where researchers discover an unexpected association in their data but report it as a prior hypothesis in the paper, as Isobel and Firenze did (Rubin, 2017; Murphy and Aguinis, 2019). In doing so, they are able to publish positive findings, even if the original hypothesis was unsupported. Rubin (2017) reports that 43 percent of researchers in recent surveys of psychologists, and some others, reported HARKing at least once (a number that may underreport its real prevalence).

HARKing is problematic for several reasons. First, in changing hypotheses, researchers often fail to report null or negative results related to the original hypothesis. This stands in the way of falsification and improperly removes disconfirmatory data from the pool of published research (Rubin, 2017). Second, if a researcher can publish on *any* association their data happen to support, they are likely to yield a positive finding, though this association is likely to be spurious. This is not necessarily problematic if the readers of the paper understands

---

[82] Although most researchers understand p-hacking as a poor practice, there is some debate on this topic. In particular, some researchers argue that it allows for discovery of unexpected findings in datasets. Hitzig and Stegenga (2020) use Bayesian models to outline when p-hacking is epistemically problematic. In particular, they argue that this is the case when it illegitimately increases the probability some evidence obtains and is updated on.

the context of the claim being made and can adjust their response accordingly. But by hiding details of the research process, authors prevent readers from properly assessing the relevance and impact of their findings.[83] Third, researchers can hide practices like p-hacking by making it seem that associations found this way were a priori (and thus not the product of multiple tests).

In fact, the conditions under which HARKing is problematic, and to what degree, are somewhat subtle.[84] Some researchers have advocated some types of HARKing as unproblematic. Rubin (2017) argues that the harm comes not in hypothesizing *after* results are known but in hypothesizing on the *basis* of known results, that is, in eliminating the independence of findings from the hypothesis.[85] Several authors use models to explore these issues. Murphy and Aguinis (2019) present a simulation that considers the effects of various sorts of HARKing on effect size estimates and knowledge accumulation. Researchers in their model are presented with sample correlations between two variables. They then choose which to report either with some HARKing method or based upon a prior hypothesis. They find that HARKing is especially pernicious when researchers also p-hack by hunting through data for the strongest possible statistical associations. This leads to overestimated effect sizes. Mohseni (2023a) uses a simple model to illustrate the conditions under which HARKing harms updated beliefs. As he points out, HARKing is only problematic when the posterior, or HARKed, hypotheses developed have a lower base rate of truth than prior hypotheses do. Thus, one problem is that scientists tend to develop good hypotheses, grounded in relevant theory, a priori. But a posteriori findings are often based on random associations and, thus, have a lower base rate of truth. On this picture, there could be cases where HARKing is actually a good thing because ex-post hypotheses are more likely to be correct than ex-ante ones. A very poor theorist, for example, may improve their study outcomes by HARKing (though this is probably unrealistic for most sciences).

6.4 Responses to the Replication Crisis

In response to the replication crisis, and especially the identification of many QRPs in real scientific communities, many scientists have proposed interventions to improve scientific standards. Here we will talk about modeling and

---

[83] Along these lines, Hollenbeck and Wright (2017) make the case that transparent HARKing (or THARKing) is unproblematic.

[84] We will not fully discuss these subtleties here.

[85] See, though, Mayo (1996) who argues that when statistical tests are "severe" enough, they will disambiguate HARKed claims that are false from those that are true regardless.

theoretical work intended to assess the potential success of these interventions. One theme that will run through this discussion is the importance of strategic considerations in thinking about intervention. Researchers respond to incentives, and, as we have seen, sometimes those responses are counterproductive. Rules intended to prevent QRPs, without changing perverse incentives may not be effective if researchers find ways around the new rules. Thus, although this section is less focused on community aspects of science, we will see that attending to these aspects is important. Regulation of science is game-theoretic – regulators respond to scientists who respond to regulators – and effective proposals for intervention should take this into account.

6.4.1 Changing the Significance Threshold

Standard frequentist statistical methods – which involve NHST – require that to test a claim, researchers must compare results to a null model on which that claim would be false. If results are unlikely given this null model, the null model is rejected. The traditional standard is to reject the null model if results cross some threshold of improbability, which has been set at 0.05. So if there is a percent chance or less that the data would be observed given the null, a finding is declared significant and often taken to support a positive claim.

One response to the replication crisis has been to propose a more stringent significance standard. In an influential, massively multiauthored paper, Benjamin et al. (2018) suggest that the new significance level should be 0.005. They point out that even in the absence of QRPs, the 0.05 threshold will mean that many stated research claims are false, and so a higher standard should improve reproducibility of results. Part of their argument involves pointing out that even if one prefers Bayesian statistics (more on these shortly), a more stringent significance threshold will tend to be associated with findings that have higher Bayes factors. They also use a simple statistical model to show how this change should reduce the false positive rate. And, lastly, they point to the relatively high reproducibility rates of psychological findings where $p < 0.005$ compared to those where $p < 0.05$.

Some theorists question the wisdom of this strategy. First, as noted, Ioannidis and others have demonstrated that significance thresholds contribute to overestimates of effect sizes, and the more stringent the threshold, the greater the expected overestimate. (This, again, is because studies that find larger effects tend to be the significant ones. Studies that meet a highly stringent threshold will tend to find highly overinflated effects.) This may negatively impact the ability of a community to correctly estimate an effect size in the presence of publication bias (Bruner and Holman, 2019). Second, Mohseni

(2023a) uses a model to illustrate how in the presence of HARKing, lowering the significance threshold might, paradoxically, lead to a higher false discovery rate. This happens when researchers engage in "fallback HARKing." They first select a hypothesis with relatively high prior probability and test it. If this test is insignificant, they scour the data for a significant association to report. These unpredicted associations have lower prior probabilities. The issue is that a stricter significance threshold means fewer researchers garner support for their original, more plausible, theses. They then go pick new hypotheses from a pool with a lower base rate of truth.

Mohseni (2023b) raises another worry, which is that a very stringent significance threshold may filter papers of different quality in different ways. Researchers using QRPs to attain significance can still use creative methods to do so, even with a more stringent requirement. Researchers adhering to high-quality practices, on the other hand, may be less likely to attain significance and, thus, less likely to publish. The result may be that stringent thresholds, paradoxically, decrease the replicability of published results because they more severely impact good researchers.

6.4.2 Reforming NHST

Others reject the suggestion to lower the significance threshold because they prefer to eliminate it altogether. McShane et al. (2019) argue that p-values should be just one tool among many – including prior evidence, plausibility of findings, and study design quality – in judging the importance and relevance of findings. As they argue, a p-value threshold for publication (1) inappropriately emphasizes specific null models of evidence, (2) arbitrarily dichotomizes a continuum of support for a hypothesis, (3) inappropriately encourages researchers to take significant findings as true ones, and vice versa, and (4) encourages QRPs designed to attain significance. In general, they encourage a wholistic approach to inference and reject the idea that "statistical alchemy" can settle the issue of whether or not an effect is of interest. These authors recognize an important issue with their proposal. As discussed throughout this Element, scientists are credit-motivated, and credit comes from attention. This directly disincentivizes any approach where scientists make muted or cautious claims on the basis of wholistic reasoning, and incentivizes approaches where meeting some arbitrary standard allows scientists to make unequivocal claims.

Another response suggests abandoning NHST altogether and adopting some version of Bayesian statistical methods (or some other alternative). In fact, many authors in the 0.005 paper prefer something along these lines but take reducing the significance threshold to be a more realistic reform. The debate over which general statistical methods are best for science is controversial and, at times, heated.[86] We do not have space to properly engage with this literature. There is a small literature, though, using simulations to compare the success of various standard inference methods over the short term.

Radzvilas et al. (2021) use simulations to argue that both Bayesianism and various sorts of frequentism all perform fairly well, and thus are all acceptable scientific tests.[87] Romero and Sprenger (2021), on the other hand, develop a model more situated in details of scientific communities, and especially precursors of the replication crisis such as publication bias. They compare NHST with graded uses of Bayes factors to label findings as more or less significant. Bayes factors are the standard Bayesian method for calculating the degree to which a set of data disambiguates between two potential hypotheses and, thus, supports some positive claim. Their model assumes that scientists draw data from a normal distribution with some prespecified effect size (or no effect). Over time, scientists publish these results, replicate them, and estimate the effect size based on meta-analysis. The authors consider both ideal conditions and cases where (1) sample sizes are small, (2) there is publication bias against null findings, and (3) there is publication bias in some direction.[88] They find that NHST tends to overestimate the effect size more often than Bayesian analysis. This is because (1) when effects are small or nonexistent, Bayesian methods do not suppress strong evidence for the null model the way NHST does and (2) when there are small sample sizes, NHST requires a very strong effect to reach the significance threshold, which skews estimates of the effect.

This provides some support for the use of Bayesian analysis in the presence of publication bias. That said, Romero and Sprenger (2021) recognize that Bayes factors can be misused and misleading as well.[89] The picture developed by McShane et al. (2019), where good inference must be wholistic and context-specific, is still a relevant one. There is no easy statistical hack that replaces careful reasoning in science.

---

[86] Radzvilas et al. (2021) describe the situation as follows: "All factions within the Statistics Wars make plausible...cases that their particular methodology, properly applied, can mitigate some of the malpractices behind the replication crisis. Furthermore, far from being dry debates, the Statistics Wars are frequently characterized by the sort of aggressive rhetoric, bombastic manifestos, and political maneuvering that their name would suggest"(13690).

[87] See also Kyburg Jr. and Teng (2013), though as Radzvilas et al. (2021) point out, there are some irregularities in their analysis.

[88] In the Bayesian version of the model, publication bias involves suppressing results with "inconclusive" Bayes factors, that is, those close to 1.

[89] See Tendeiro and Kiers (2019) and Mayo (2018), for example.

6.4.3 Registered Reports, Journals for Null Results, and Preregistration

One way to reduce the perverse incentives created by publication bias is to actually publish null or negative results. In this way, all findings receive credit, and there is less incentive to commit fraud or to use QRPs. In addition, this creates incentives for replication work.

Along these lines, some journals now use *registered reports*, which are submitted before research is conducted (Chambers, 2013; Nosek and Lakens, 2014). Reviewers judge the methodological quality of the proposed research. On the grounds of this review, journals decide whether to publish the work, regardless of the study outcome. Scientists then perform the research and write up their results. As long as their methods adhere to the registered report, their fundings are published. Allen and Mehler (2019) found that registered reports showed a much larger publication rate for null findings than the general literature. Recall from Section 2 that Gross and Bergstrom (2021) raise worries that pre-review of this sort will contribute to conservatism in science as authors attempt to please the average reviewer. But if this review is based on methodological quality, rather than specific hypotheses tested, this is less of a worry.

An alternative approach is the creation of journals and venues to publish and share null and negative results (Munafò et al., 2017). Current examples include *PLOS One*'s "Missing Pieces" collection or the *Journal of Negative Results in Biomedicine*. Both these reforms should, ideally, lessen the negative impacts of publication bias. Nosek et al. (2012), though, argue that such venues are relatively ineffective as they just further enshrine the low-status associated with null findings. Thus, they may not generate credit for null results in the right way.

Registered reports ensure publication of negative or null results, but they also play another important role in constraining research practices before they happen. Preregistration is a widely deployed intervention that does only the latter (Nosek et al., 2018; Lakens, 2019). Researchers submit detailed research plans to open access sites, like Open Science Foundation. They then perform their research as planned and include their preregistration in the publication. This reduces researcher degrees of freedom by preventing p-hacking, HARKing, and forking paths. When study changes are necessary, preregistration requires transparency with respect to what was changed and why. Proponents point out that preregistration in clinical studies drastically reduced reported positive outcomes (Kaplan and Irvin, 2015).

There have been a number of criticisms of preregistration, though, including worries that it will prevent exploratory research, or novel discovery, and that it still allows researcher degrees of freedom but just creates a veneer of credibility. (I do not overview this literature, but see Szollosi et al. (2020) and Rubin (2020), for examples.) For this reason, Hitzig and Stegenga (2020) argue for flexible preregistration. Scientists can deviate from research plans but, in doing so, must be transparent about how and why they do so.

Mohseni (2023b) models the success of interventions like preregistration in the social context of scientific communities. As he points out, there may be ways that preregistration could lead to responses that are counterproductive. For instance, if researchers choose to preregister only in cases where they are highly confident of results, but not other cases, the intervention may end up constraining only those with the highest base rates of true hypotheses, thus decreasing the reliability of findings overall. But, note that his analysis applies only to optional preregistration. If preregistration is a generally required standard, this worry does not stand.

6.4.4 Transparency in Practice

A general trend intended to address the replication crisis is to improve transparency in scientific practice (Schofield et al., 2009; Nosek et al., 2012; Nosek and Bar-Anan, 2012). Preregistration does this, and so do initiatives to require the sharing of data, rather than just analyzed results. Many journals now require authors to publish their data, code, or other materials on a forum such as open science foundation (OSF). This practice means that other researchers can potentially identify errors, fraud, or deviations from preregistration.

This sort of transparency is also related to the Open Science movement, which aims to ensure scientific data and findings are shared as widely as possible. This initiative is intended to improve equity and also to more perfectly align practice with the communist norm. The mushrooming of preprint servers, where authors can share early drafts as well as papers ready for publication, reflects this movement. One additional benefit is that authors can get critical feedback from a wider range of academics, and earlier in the research process (Bourne et al., 2017).

While such reforms are widely supported, one small worry relates to results described in Section 4 about sharing and transient diversity. One way to ensure diversity of practice is to preserve a diversity of beliefs. If data and findings are shared too widely and quickly, this may sometimes reduce such diversity.[90] On balance, though, the benefits of communism of this sort likely outweigh this cost.

---

[90] Thanks to Kaetlin Taylor for this point.

6.4.5 Improving Theory

As noted, the false discovery rate of a discipline is strongly shaped by the base rate of true hypotheses tested. For this reason, some theorists have suggested that important interventions should come long before we start constraining the flexibility of methods or improving statistical tests. In particular, science should take steps to ensure that researchers are testing more, better hypotheses whenever possible.

In some cases, such as in highly exploratory research, this will not be possible. In these areas, there is not enough theoretical understanding of a phenomenon to constrain hypotheses to just the good ones. In other areas, such as biomedical research, there are so many possible hypotheses to test that it is not possible to create very high base rates.

When possible, though, Smaldino (2019a) argues that good theory is necessary to improve the base rate of true hypotheses in scientific disciplines. As he points out, studies in cognitive psychology have replicated at about twice the rate of those in social psychology, arguably because they rely on more solid theoretical grounds (Open Science et al., 2015). Without good theory, hypotheses are relatively unconstrained and need not fit with our best understandings of the workings of the world.[91] Note that this particular intervention fits with current incentives. Researchers want to generate accurate hypotheses in order to publish.

6.4.6 Multiple Labs and Adversarial Collaborations

In adversarial collaborations, multiple scientists or teams that disagree on some hypothesis work together to develop a research program designed to test it (Latham et al., 1988; Nuzzo, 2015). The idea is to avoid the psychological tendencies that might push a researcher toward choices that end up confirming their beliefs. Alternatively, in multiple-lab approaches, multiple teams who are not necessarily adversarial all work together to conduct the same experiments. They collaborate to ensure uniformity of approaches and then are able to assess whether results are similar during a single time slice. The "many labs" project used this technique to assess replicability of previous findings (Klein et al., 2014, 2018). Protzko et al. (2023) use the technique to show how, with good enough methodology, novel findings in the social sciences can be highly replicable. Although multiple-lab approaches are more difficult and expensive

---

[91] Aydin Mohseni, in personal correspondence, has also argued that good theory also acts similarly to preregistration, in that it prevents HARKing. If, say, one tests a game-theoretic prediction that turns out false, one cannot easily claim that their initial hypothesis was that this overarching behavioral theory would be wrong.

to coordinate and carry out than standard research, they do point to a way to improve methodological quality in areas where replicability has been an issue.

• • •

Taken together, we can see that modeling work is playing a core role in both (1) identifying causes of the replication crisis and (2) exploring the costs and benefits of various interventions meant to improve the quality of research. This work extends purely theoretical arguments by adding logical constraints to reasoning about them. While it is usually problematic to go directly from modeling work to policy interventions, the models here are extremely useful in identifying possible outcomes of various interventions and directing future empirical work on the best structures for science. In this way, they are important tools in the metascience toolkit. Attempts to intervene on scientific communities do well to draw on both models and other sorts of work in identifying the best ways to do so.

7 Conclusion

This Element overviewed a wide range of models intended to illuminate the workings of scientific communities. As we have seen, there are a number of formalisms and approaches that help do this. Credit-economy models consider the impacts of incentive structures on the first-order decision-making of scientists and other agents in science, such as journals and grant-giving bodies. Selective models consider how forces related to persistence and copying in scientific communities can shape methods and practices. Instead of considering the effects of credit on decision-making, they consider the effects of credit on which sorts of practices are replicated and who remains in a community. Network models spotlight communication and the way that beliefs and ideas spread in epistemic communities. This spread, in turn, impacts research choices and thus the progress of inquiry. Epistemic landscape models give an insight into problem choice and how a host of factors – personalities of scientists, grant-giving policy, and community interactions – shape what sorts of topics get explored in science. And models of statistical practice have contributed to explorations of the replication crisis, QRPs, improving practice, and metascience more generally.

In concluding this Element, I will do two things. First, I think it will be helpful to overview some of the policy suggestions described in this Element. Many of the suggestions across different models dovetail or interrelate, and it is worth discussing them all together and pulling out themes. Second, I want to return to the question of what we should take away from models of the sort described, especially with regard to policy.

Probably the strongest policy theme throughout the Element focused on improving the quality of scientific methods and the replicability of findings. In this vein, in Sections 2 and 3, we saw various proposals aimed at increasing replication work, with the goal of better testing the replicability of findings (Stewart and Plotkin, 2021). These involved requiring replications in papers reporting new findings (Begley and Ellis, 2012) and creating communities of scientists incentivized only to do replication work (Romero, 2018, 2020). Related to this were proposals to improve incentives for fraud detection since these incentives should, likewise, improve scientific self-policing (Bruner, 2013). Proposals from Section 6 regarding transparency in science dovetail with these proposals (Schofield et al., 2009; Nosek and Bar-Anan, 2012; Nosek et al., 2012). Transparency in practice allows other researchers to assess exactly what has been done, check for signs of fraud, and successfully repeat experiments. One worry about these proposals is that replications take a lot of work. It is not particularly efficient to run many low-quality studies and then repeat them many times in efforts to discover which findings are good ones.

Alternatively, we saw many proposals, especially in Sections 3 and 6, aimed at improving the quality of studies in the first place. These interventions included improving theory to increase the base rate of true hypotheses tested (Smaldino, 2019a; Bird, 2021; Stewart and Plotkin, 2021), changing statistical practices (such as switching to Bayesianism or lowering the significance threshold) (Benjamin et al., 2018; McShane et al., 2019; Romero and Sprenger, 2021), requiring preregistration (Nosek et al., 2018; Lakens, 2019), and moving toward more many-lab collaborations.[92] On balance of evidence, improving the base rate of true hypotheses (wherever possible) seems an uncontroversially good idea. The other policies have some possible downsides such as time/money costs (many-labs) and restrictions to exploration (preregistration), though natural tests of them are already underway as some disciplines adopt these practices. Changes to statistical standards are, likewise, worth testing further.

Many of the proposals just listed are intended to reverse the effects of perverse credit incentives. As long as scientists get credit for quickly making positive discoveries, they will be incentivized to use practices that quickly make positive discoveries. As we saw, this can drive scientists to choose sloppy, questionable, or fraudulent practices, and it can also drive selection pressures for low-quality practices (Higginson and Munafò, 2016; Smaldino and McElreath, 2016; Bright, 2017b; Heesen, 2018, 2021; Zollman, 2023). The issue,

---

[92] In addition, Smaldino and O'Connor (2022) argue for interdisciplinary contact as a general mechanism to promote the spread of high-quality methods.

of course, is that many of these practices are not particularly truth-conducive. This might make us think that rather than committing a lot of time and energy to replication, or implementing post-hoc barriers that prevent the publication of poor work, we ought to change the incentive structure that drives it.

This insight seems especially germane, given a few things. First, we saw models illustrating that even with a lot of replication, we can fail to form good beliefs in the presence of publication bias (Nissen et al., 2016; Romero, 2016; Devezer et al., 2019). This happens when only positive findings are published, rather than a representative sample of studies. If so, attempts to increase replication may be not just inefficient but unsuccessful. And, second, we saw general worries that as long as perverse incentives are in place, scientists will find creative ways to get around interventions aimed to improve the quality of work (Mohseni, 2023a, 2023b).

Some of the policy proposals described are oriented toward changing credit-incentive structures themselves. Registered reports, for example, help incentivize the production of high-quality research (rather than positive results) (Chambers, 2013; Nosek and Lakens, 2014). And journals for null and negative results may do likewise (Munafò et al., 2017). Adversarial collaborations also change incentive schemes, since they draw attention to work developing the highest-quality tests possible (Latham et al., 1988; Nuzzo, 2015). Scoop protection is meant to protect slow, high-quality work from priority incentives (Tiokhin et al., 2021). Transparency policies arguably change incentives as well – typically in order to receive credit, scientists must at least be perceived as using high-quality methods. Interventions that make methodological quality easier to observe put pressure on researchers to do good work (Smaldino et al., 2019).

There are tensions regarding attempts to change credit incentives, though. First, credit is not assigned purely as a result of science policy. As we have seen, it often accrues piecemeal, though a process of citation, attribution, and response from a community. In particular, credit in part derives from human preferences for novelty. Although we can create journals for null results, we cannot make those results exciting. Second, as discussed in Section 2, credit structures like the priority rule play significant roles in promoting communism and productivity, both of which are crucial to scientific progress. That said, eliminating publication bias and significance thresholds should preserve positive impacts of the priority rule, while removing some of the perverse incentives to generate positive findings.

Another policy theme we saw throughout the Element, and especially in Sections 4 and 5, is related to the promotion of transient diversity of practice. In order to function well, scientific communities must engage in some significant amount of exploration. The question is: How do we promote this exploration?

In Section 4, we saw models pointing to potential benefits for decreased communication in science (Zollman, 2010). If scientists communicate less, they might maintain a greater diversity of opinions, assumptions, and interests. While this particular proposal likely has risky downsides, there are other ways to yield similar positive effects. The use of centralized funding bodies to coordinate exploration across research groups can promote diversity of practice without creating inefficiencies from cutting off information flow (Goldman, 1999; Viola, 2015; Kummerfeld and Zollman, 2020; Wu and O'Connor, 2023). Relatedly, Zollman (2009) argues for journal strategies to share diverse findings in order to promote transient diversity, especially by using randomization to select articles to publish. And in Section 5, Avin (2019) and Harnagel (2019) argued that random elements in funding schemes might play the same role as highly exploratory strategies and might be beneficial for that reason. (Funding lotteries were also supported elsewhere in the Element for their benefits to efficiency and to the natural selection of methods. Altogether, they seem to be promising for a number of reasons.)

Several models indicate that beneficial diversity of practice can be generated by including different sorts of people in science. Demographic diversity in network models can act as a driver of transient diversity of practice and improved learning (Fazelpour and Steel, 2022; Wu, 2023b). And whatever policies promote cognitive diversity more generally may be helpful in the light of landscape model results, indicating that individuals with different cognitive styles may more successfully search a landscape of ideas (Thoma, 2015; Pöyhönen, 2017; Devezer et al., 2019; Hong and Page, 2004). With respect to which of these mechanisms is best for encouraging transient diversity in science, promoting demographic diversity seems like an obvious win. It is worthwhile addressing historical inequities in any case, and there are few downside risks to this sort of intervention.

Another policy theme that ran throughout the Element is related to increasing the efficiency of scientific communities. Whatever time scientists waste on reviewing, writing grant proposals, and so on is time they do not spend making important discoveries. Gross and Bergstrom (2019) argued for grant-reviewing systems that partially depend on lotteries to settle funding. This might save time for both reviewers and those submitting grants. Likewise Heesen and Bright (2020) and Arvan et al. (2020) argue for ending prepublication peer review entirely, in part for efficiency reasons. Doing only postpublication review would speed up the spread of new findings and save time spent by reviewers, editors, and authors. While both these proposals seem promising, there are potential downsides (regarding social influences, conservatism, and inequity), and testing before widespread implementation seems merited. Such tests are already naturally underway as just some grant bodies, some journals, and some disciplines experiment with new reviewing structures.

One last theme, especially present in Section 4, regarded responses to industrial influence on science. Holman and Bruner (2015), Lewandowsky et al. (2019), and Weatherall et al. (2020) all suggest that industrial influence strategies ranging from data fabrication to cherry-picking can work to sway scientific consensus and public understanding. On the basis of this work, O'Connor and Weatherall (2019) argue for attempts to remove industry influence from science, perhaps by funneling industry research money through centralized agencies that shape who does the research and how. Weatherall et al. (2020) argue for quality checks on published research to prevent cherry-picking. And Holman and Bruner (2015) recommend methods of detection for industrial actors. Given the massive historical harms industry has had on scientific progress and public belief, these types of interventions warrant further investigation. In general, metascience and philosophy of science should be more focused on reinventing funding structures to protect science from industry harms.

Wu (2023a) recall focuses on another aspect of industry science – that industry can benefit from the communist norm without, in turn, having to share their own research. On the basis of this research, Wu and O'Connor (2023) advocate for required sharing of industry research. This policy obviously needs further testing and assessment, but it may have significant benefits to scientific progress.

Throughout the Element, the models reviewed are mostly simple. They idealize and abstract away from details of the communities they represent. This allows for tractability. Many of these models also helpfully illuminate causes and effects in scientific communities in ways that might not be possible in more complicated models. But, as always, simple models must be treated with care. They should not be taken to simply reveal truths about how science works. That said, the models overviewed *can* successfully play many and varied roles in inquiry into the workings of science. They can help researchers explore possibilities, identify plausible causes of various maladies, sort out what certain interventions might do, suggest further empirical work, cast doubt on theoretical claims, act as more constrained versions of thought-experiments or theoretical reasoning, and so on. We have seen models play all these sorts of roles throughout this Element. Models are an important set of tools in our

attempts to understand and improve the workings of scientific communities. As
outlined in the introduction, much of the time, they are best used in conjunc-
tion with empirical tools to improve our understanding of science. And as we
have seen, in many cases, this is exactly what is happening in various scientific
disciplines as they test interventions or policies intended to improve scientific
practice.

References

Akerlof, George A and Pascal Michaillat (2018). "Persistence of false paradigms in low-power sciences." *Proceedings of the National Academy of Sciences*, *115*(52), 13228–13233.

Alexander, Diane, Olga Gorelkina, and Erin Hengel (2021). "Gender and the time cost of peer review." Working Paper.

Alexander, Jason McKenzie, Johannes Himmelreich, and Christopher Thompson (2015). "Epistemic landscapes, optimal search, and the division of cognitive labor." *Philosophy of Science*, *82*(3), 424–453.

Allen, Christopher and David MA Mehler (2019). "Open science challenges, benefits and tips in early career and beyond." *Public Library of Science Biology*, *17*(5), e3000246.

Anderson, Katharine A (2016). "A model of collaboration network formation with heterogeneous skills." *Network Science*, *4*(2), 188–215.

Arvan, Marcus, Liam Kofi Bright, and Remco Heesen (2020). "Jury theorems for peer review." *The British Journal for the Philosophy of Science*. https://doi.org/10.1086/719117.

Avin, Shahar (2015). "Funding science by lottery." *Recent Developments in the Philosophy of Science: EPSA13 Helsinki*. Springer, 111–126.

Avin, Shahar (2019). "Centralized funding and epistemic exploration." *The British Journal for the Philosophy of Science*, *70*(3), 629–656.

Azar, Ofer H (2005). "The review process in economics: Is it too fast?" *Southern Economic Journal*, *72*(2), 482–491.

Baker, Monya (2016). "1,500 scientists lift the lid on reproducibility." *Nature News*, *533*(7604), 452.

Bala, Venkatesh and Sanjeev Goyal (1998). "Learning from neighbours." *The Review of Economic Studies*, *65*(3), 595–621.

Balietti, Stefano, Michael Mäs, and Dirk Helbing (2015). "On disciplinary fragmentation and scientific progress." *Public Library of Science One*, *10*(3), e0118747.

Banerjee, Siddhartha, Ashish Goel, and Anilesh Kollagunta Krishnaswamy (2014). "Re-incentivizing discovery: Mechanisms for partial-progress sharing in research." *Proceedings of the Fifteenth ACM Conference on Economics and Computation*, 149–166.

Barabâsi, Albert-Laszlo, Hawoong Jeong, Zoltan Néda, et al. (2002). "Evolution of the social network of scientific collaborations." *Physica A: Statistical Mechanics and Its Applications*, *311*(3–4), 590–614.

Barkoczi, Daniel and Mirta Galesic (2016). "Social learning strategies modify the effect of network structure on group performance." *Nature Communications*, *7*(1), 1–8.

Bauchner, Howard, Phil B Fontanarosa, Annette Flanagin, and Joe Thornton (2018). "Scientific misconduct and medical journals." *Jama*, *320*(19), 1985–1987.

Bedessem, Baptiste (2019). "The division of cognitive labor: Two missing dimensions of the debate." *European Journal for Philosophy of Science*, *9*(1), 1–16.

Begley, C Glenn and Lee M Ellis (2012). "Raise standards for preclinical cancer research." *Nature*, *483*(7391), 531–533.

Bender, Max Ernst, Suzanne Edwards, Peter von Philipsborn, et al. (2015). "Using co-authorship networks to map and analyse global neglected tropical disease research with an affiliation to Germany." *Public Library of Science Neglected Tropical Diseases*, *9*(12), e0004182.

Benjamin, Daniel J, James O Berger, Magnus Johannesson, et al. (2018). "Redefine statistical significance." *Nature Human Behaviour*, *2*(1), 6–10.

Bergstrom, Carl T, Jacob G Foster, and Yangbo Song (2016). "Why scientists chase big problems: Individual strategy and social optimality." *ArXiv Preprint ArXiv:1605.05822*.

Bird, Alexander (2021). "Understanding the replication crisis as a base rate fallacy." *The British Journal for the Philosophy of Science*, *72*(4), 965–993.

Boroomand, Amin and Paul E Smaldino (2021). "Hard work, risk-taking, and diversity in a model of collective problem solving." *Journal of Artificial Societies and Social Simulation*, *24*(4) 10.

Boschini, Anne and Anna Sjögren (2007). "Is team formation gender neutral? Evidence from coauthorship patterns." *Journal of Labor Economics*, *25*(2), 325–365.

Botts, Tina Fernandes, Liam Kofi Bright, Myisha Cherry, Guntur Mallarangeng, and Quayshawn Spencer (2014). "What is the state of blacks in philosophy?" *Critical Philosophy of Race*, *2*(2), 224–242.

Bourne, Philip E, Jessica K Polka, Ronald D Vale, and Robert Kiley (2017). "Ten simple rules to consider regarding preprint submission." *Public Library of Science Computational Biology*, *13*, e1005473.

Boyer, Thomas (2014). "Is a bird in the hand worth two in the bush? Or, whether scientists should publish intermediate results." *Synthese*, *191*(1), 17–35.

Boyer-Kassem, Thomas and Cyrille Imbert (2015). "Scientific collaboration: Do two heads need to be more than twice better than one?" *Philosophy of Science*, *82*(4), 667–688.

Bright, Liam K (2017a). "Decision theoretic model of the productivity gap." *Erkenntnis*, *82*(2), 421–442.

Bright, Liam K (2017b). "On fraud." *Philosophical Studies*, *174*(2), 291–310.

Bright, Liam K (2021). "Why do scientists lie?" *Royal Institute of Philosophy Supplements*, *89*, 117–129.

Bright, Liam K, Haixin Dang, and Remco Heesen (2018). "A role for judgment aggregation in coauthoring scientific papers." *Erkenntnis*, *83*(2), 231–252.

Bruner, Justin and Cailin O'Connor (2017). "Power, bargaining, and collaboration." In T. Boyer-Kassem, C. Mayo-Wilson, and M. Weisberg (eds.), *Scientific Collaboration and Collective Knowledge*. Oxford University Press, 135–160.

Bruner, Justin P (2013). "Policing epistemic communities." *Episteme*, *10*(4), 403–416.

Bruner, Justin P (2019). "Minority (dis) advantage in population games." *Synthese*, *196*(1), 413–427.

Bruner, Justin P and Bennett Holman (2019). "Self-correction in science: Meta-analysis, bias and social structure." *Studies in History and Philosophy of Science Part A*, *78*, 93–97.

Camerer, Colin F, Anna Dreber, Felix Holzmeister, et al. (2018). "Evaluating the replicability of social science experiments in Nature and Science between 2010 and 2015." *Nature Human Behaviour*, *2*(9), 637–644.

Campbell, Donald T (1965). "Variation and selective retention in socio-cultural evolution." *Social Change in Developing Area*.

Casadevall, Arturo and Ferric C Fang (2012). "Reforming science: Methodological and cultural reforms." *Infection and Immunity*, *80*, 891–896.

Chambers, Christopher D (2013). "Registered reports: A new publishing initiative at Cortex." *Cortex*, *49*(3), 609–610.

Collins, Harry M (1974). "The TEA set: Tacit knowledge and scientific networks." *Science Studies*, *4*(2), 165–185.

Cotton, Christopher (2013). "Submission fees and response times in academic publishing." *American Economic Review*, *103*(1), 501–509.

Currie, Adrian (2019). "Existential risk, creativity & well-adapted science." *Studies in History and Philosophy of Science Part A*, *76*, 39–48.

Dahlberg, Brett (2018). "Cornell food researcher's downfall raises larger questions for science." *NPR*.

Dasgupta, Partha and Paul A David (1994). "Toward a new economics of science." *Research Policy*, *23*(5), 487–521.

Dasgupta, Partha and Eric Maskin (1987). "The simple economics of research portfolios." *The Economic Journal*, *97*(387), 581–595.

De Langhe, Rogier (2014). "A unified model of the division of cognitive labor." *Philosophy of Science*, *81*(3), 444–459.

de Melo-Martín, Inmaculadade and Kristen Intemann (2018). *The fight against doubt: How to bridge the gap between scientists and the public*. Oxford University Press.

Del Carmen, Alejandro and Robert L Bing (2000). "Academic productivity of African Americans in criminology and criminal justice." *Journal of Criminal Justice Education*, *11*(2), 237–249.

Derex, Maxime, Charles Perreault, and Robert Boyd (2018). "Divide and conquer: Intermediate levels of population fragmentation maximize cultural accumulation." *Philosophical Transactions of the Royal Society B: Biological Sciences*, *373*(1743), 20170062.

Devezer, Berna, Luis G Nardin, Bert Baumgaertner, and Erkan Ozge Buzbas (2019). "Scientific discovery in a model-centric framework: Reproducibility, innovation, and epistemic diversity." *Public Library of Science One*, *14*(5), e0216125.

Devezer, Berna, Danielle J Navarro, Joachim Vandekerckhove, and Erkan Ozge Buzbas (2021). "The case for formal methodology in scientific reform." *Royal Society Open Science*, *8*(3), 200805.

Dion, Michelle L, Jane Lawrence Sumner, and Sara McLaughlin Mitchell (2018). "Gendered citation patterns across political science and social science methodology fields." *Political Analysis*, *26*(3), 312–327.

Dotson, Kristie (2011). "Tracking epistemic violence, tracking practices of silencing." *Hypatia*, *26*(2), 236–257.

Douglas, Heather, Kevin Elliott, Andrew Maynard, Paul Thompson, and Kyle Whyte (2014). "Guidance on Funding from Industry." *SRPoiSE.org*. http://srpoise.org/wp-content/uploads/2014/06/Guidance-on-Funding-from-Industry-Final.pdf.

Du Bois, WEB (1898). "The study of the Negro problems." *The Annals of the American Academy of Political and Social Science*, *11*(1), 1–23.

Eklund, Anders, Thomas E Nichols, and Hans Knutsson (2016). "Cluster failure: Why fMRI inferences for spatial extent have inflated false-positive rates." *Proceedings of the National Academy of Sciences*, *113*(28), 7900–7905.

Etzkowitz, Henry, Stephan Fuchs, Namrata Gupta, Carol Kemelgor, and Marina Ranga (2008). "The coming gender revolution in science." In Edward J. Hackett, Olga Amsterdamska, Michael Lynch, and Judy Wajcman (eds.), *The Handbook of Science and Technology Studies*. Cambridge: MIT Press, 403–428.

Fanelli, Daniele (2009). "How many scientists fabricate and falsify research? A systematic review and meta-analysis of survey data." *Public Library of Science One*, 4(5), e5738.

Fanelli, Daniele (2012). "Negative results are disappearing from most disciplines and countries." *Scientometrics*, 90(3), 891–904.

Fang, Christina, Jeho Lee, and Melissa A Schilling (2010). "Balancing exploration and exploitation through structural design: The isolation of subgroups and organizational learning." *Organization Science*, 21(3), 625–642.

Fazelpour, Sina and Hannah Rubin (2022). "Diversity and homophily in social networks." *Proceedings of the Annual Meeting of the Cognitive Science Society*.

Fazelpour, Sina and Daniel Steel (2022). "Diversity, trust and conformity: A simulation study." *Philosophy of Science*, 89(2), 209–231.

Feldon, David F, James Peugh, Michelle A Maher, Josipa Roksa, and Colby Tofel-Grehl (2017). "Time-to-credit gender inequities of first-year PhD students in the biological sciences." *CBE Life Sciences Education*, 16(1), 1–9.

Ferber, Marianne A and Michael Brün (2011). "The gender gap in citations: Does it persist?" *Feminist Economics*, 17(1), 151–158.

Ferber, Marianne A and Michelle Teiman (1980). "Are women economists at a disadvantage in publishing journal articles?" *Eastern Economic Journal*, 6(3/4), 189–193.

Frey, Daniel and Dunja Šešelja (2018). "What is the epistemic function of highly idealized agent-based models of scientific inquiry?" *Philosophy of the Social Sciences*, 48(4), 407–433.

Frey, Daniel and Dunja Šešelja (2020). "Robustness and idealizations in agent-based models of scientific interaction." *The British Journal for the Philosophy of Science*, 71(4), 1411-1437.

Fricker, Miranda (2007). *Epistemic Injustice: Power and the Ethics of Knowing*. Oxford University Press.

Gabriel, Nathan and Cailin O'Connor (2023). "Can confirmation bias improve group learning? Working Paper.

Gadbury, Gary L and David B Allison (2012). "Inappropriate fiddling with statistical analyses to obtain a desirable p-value: Tests to detect its presence in published literature." *Plos One*, 7(1), e46363.

Gelman, Andrew and Eric Loken (2013). "The garden of forking paths: Why multiple comparisons can be a problem, even when there is no 'fishing expedition' or 'p-hacking' and the research hypothesis was posited ahead of time." *Department of Statistics, Columbia University, 348*, 1–17.

Glänzel, Wolfgang and András Schubert (2004). "Analysing scientific networks through co-authorship." In H. Moed, W. Glanzel, U. Smoch (eds.), *Handbook of Quantitative Science and Technology Research*. Springer, 257–276.

Goldman, Alvin I (1999). *Knowledge in a Social World*. Oxford University Press.

Goldman, Alvin I and Moshe Shaked (1991). "An economic model of scientific activity and truth acquisition." *Philosophical Studies: An International Journal for Philosophy in the Analytic Tradition*, *63*(1), 31–55.

Golub, Benjamin and Matthew O Jackson (2010). "Naive learning in social networks and the wisdom of crowds." *American Economic Journal: Microeconomics*, *2*(1), 112–149.

Golub, Benjamin and Matthew O Jackson (2012). "How homophily affects the speed of learning and best-response dynamics." *The Quarterly Journal of Economics*, *127*(3), 1287–1338.

Goodman, Steven and Sander Greenland (2007). "Why most published research findings are false: Problems in the analysis." *Public Library of Science Medicine*, *4*(4), e168.

Gopalakrishna, Gowri, Gerben Ter Riet, Gerko Vink, et al. (2022). "Prevalence of questionable research practices, research misconduct and their potential explanatory factors: A survey among academic researchers in the Netherlands." *Public Library of Science One*, *17*(2), e0263023.

Gorman, Dennis M, Amber D Elkins, and Mark Lawley (2019). "A systems approach to understanding and improving research integrity." *Science and Engineering Ethics*, *25*(1), 211–229.

Grim, Patrick, Daniel J Singer, Aaron Bramson, et al. (2019). "Diversity, ability, and expertise in epistemic communities." *Philosophy of Science*, *86*(1), 98–123.

Grim, Patrick, Daniel J Singer, Steven Fisher, et al. (2013). "Scientific networks on data landscapes: Question difficulty, epistemic success, and convergence." *Episteme*, *10*(4), 441–464.

Grim, Patrick, Daniel J Singer, Christopher Reade, and Steven Fisher (2015). "Germs, genes, and memes: Function and fitness dynamics on information networks." *Philosophy of Science*, *82*(2), 219–243.

Gross, Kevin and Carl T Bergstrom (2019). "Contest models highlight inherent inefficiencies of scientific funding competitions." *Public Library of Science Biology*, *17*(1), e3000065.

Gross, Kevin and Carl T Bergstrom (2021). "Why ex post peer review encourages high-risk research while ex ante review discourages it." *Proceedings of the National Academy of Sciences*, *118*(51), e2111615118.

Harnagel, Audrey (2019). "A mid-level approach to modeling scientific communities." *Studies in History and Philosophy of Science Part A, 76,* 49–59.

Head, Megan L, Luke Holman, Rob Lanfear, Andrew T Kahn, and Michael D Jennions (2015). "The extent and consequences of p-hacking in science." *Public Library of Science Biology, 13*(3), e1002106.

Heesen, Remco (2017a). "Academic superstars: Competent or lucky?" *Synthese, 194*(11), 4499–4518.

Heesen, Remco (2017b). "Communism and the incentive to share in science." *Philosophy of Science, 84*(4), 698–716.

Heesen, Remco (2018). "Why the reward structure of science makes reproducibility problems inevitable." *The Journal of Philosophy, 115*(12), 661–674.

Heesen, Remco (2021). "Cumulative advantage and the incentive to commit fraud in science." *Studies in History and Philosophy of Science Part A.* https://doi.org/10.1086/716235.

Heesen, Remco and Liam Kofi Bright (2020). "Is peer review a good idea?" *The British Journal for the Philosophy of Science, 72*(3), 635–663.

Heesen, Remco, Liam Kofi Bright, and Andrew Zucker (2019). "Vindicating methodological triangulation." *Synthese, 196*(8), 3067–3081.

Heesen, Remco and Jan-Willem Romeijn (2019). "Epistemic diversity and editor decisions: A statistical Matthew effect." *Philosophers' Imprint, 19*(39), 1–20.

Hegselmann, Rainer, Ulrich Krause (2002). "Opinion dynamics and bounded confidence models, analysis, and simulation." *Journal of Artificial Societies and Social Simulation, 5*(3) 1–33.

Hengel, Erin (2022). "Publishing while female: Are women held to higher standards? Evidence from peer review." *The Economic Journal, 132*(648), 2951–2991. https://doi.org/10.1093/ej/ueac032/6586337.

Higginson, Andrew D and Marcus R Munafò (2016). "Current incentives for scientists lead to underpowered studies with erroneous conclusions." *Public Library of Science Biology, 14*(11), e2000995.

Hightower, Jane Marie (2011). *Diagnosis: Mercury: Money, Politics, and Poison.* Island Press.

Hitzig, Zoe and Jacob Stegenga (2020). "The problem of new evidence: P-hacking and pre-analysis plans." *Diametros, 17*(66), 1–24.

Hollenbeck, John R and Patrick M Wright (2017). "Harking, sharking, and tharking: Making the case for post hoc analysis of scientific data." *Journal of Management, 43,* 5–18.

Holman, Bennett and Justin Bruner (2017). "Experimentation by industrial selection." *Philosophy of Science*, *84*(5), 1008–1019.

Holman, Bennett and Justin P Bruner (2015). "The problem of intransigently biased agents." *Philosophy of Science*, *82*(5), 956–968.

Hong, Lu and Scott E Page (2004). "Groups of diverse problem solvers can outperform groups of high-ability problem solvers." *Proceedings of the National Academy of Sciences*, *101*(46), 16385–16389.

Huebner, Bryce and Liam Kofi Bright (2020). "Collective responsibility and fraud in scientific communities." In S. Bazargan-Forward and D. P. Tollefsen (eds.), *The Routledge Handbook of Collective Responsibility*, 358–372.

Hull, David L (1988). Science as a Process: An Evolutionary Account of the Social and Conceptual Development of Science. University of Chicago Press.

Ioannidis, John PA (2005). "Why most published research findings are false." *Public Library of Science Medicine*, *2*(8), e124.

Ioannidis, John PA (2008). "Why most discovered true associations are inflated." *Epidemiology*, *19*(5), 640–648.

Jackson, Matthew O and Asher Wolinsky (2003). *A Strategic Model of Social and Economic Networks*. Springer.

John, Leslie K, George Loewenstein, and Drazen Prelec (2012). "Measuring the prevalence of questionable research practices with incentives for truth telling." *Psychological Science*, *23*(5), 524–532.

Jönsson, Martin L, Ulrike Hahn, and Erik J Olsson (2015). "The kind of group you want to belong to: Effects of group structure on group accuracy." *Cognition*, *142*, 191–204.

Kaplan, Robert M and Veronica L Irvin (2015). "Likelihood of null effects of large NHLBI clinical trials has increased over time." *Public Library of Science One*, *10*(8), e0132382.

Kauffman, Stuart and Simon Levin (1987). "Towards a general theory of adaptive walks on rugged landscapes." *Journal of Theoretical Biology*, *128*(1), 11–45.

Kauffman, Stuart A and Edward D Weinberger (1989). "The NK model of rugged fitness landscapes and its application to maturation of the immune response." *Journal of Theoretical Biology*, *141*(2), 211–245.

Kerr, Norbert L (1998). "HARKing: Hypothesizing after the results are known." *Personality and Social Psychology Review*, *2*(3), 196–217.

Kitcher, Philip (1990). "The division of cognitive labor." *The Journal of Philosophy*, *87*(1), 5–22.

Klein, Richard A, Kate A Ratliff, Michelangelo Vianello, et al. (2014). "Investigating variation in replicability: A 'many labs' replication project." *Social Psychology*, *45*(3), 142–152.

Klein, Richard A, Michelangelo Vianello, Fred Hasselman, et al. (2018). "Many labs 2: Investigating variation in replicability across samples and settings." *Advances in Methods and Practices in Psychological Science*, *1*(4), 443–490.

Kleinberg, Jon and Sigal Oren (2011). "Mechanisms for (mis) allocating scientific credit." *Proceedings of the Forty-Third Annual ACM Symposium on Theory of Computing*. 529–538.

Korf, Rebecca (2023). Taking the Social Structure of Science Seriously When Debating Values in Science. Working Paper.

Kuhn, Thomas (1962). *The Structure of Scientific Revolutions*. Princeton University Press.

Kummerfeld, Erich and Kevin JS Zollman (2020). "Conservatism and the scientific state of nature." *The British Journal for the Philosophy of Science*, *67*(4) 1057–1076.

Kyburg Jr, Henry E and Choh Man Teng (2013). "Choosing among interpretations of probability." *ArXiv Preprint ArXiv:1301.6713*.

LaCroix, Travis, Anders Geil, and Cailin O'Connor (2021). "The dynamics of retraction in epistemic networks." *Philosophy of Science*, *88*(3), 415–438.

Lakens, Daniel (2019). "The value of preregistration for psychological science: A conceptual analysis." *Japanese Psychological Review*, *62*(3), 221–230.

Landemore, Hélène (2012). *Democratic Reason*. Princeton University Press.

Larivière, Vincent, Chaoqun Ni, Yves Gingras, Blaise Cronin, and Cassidy R Sugimoto (2013). "Bibliometrics: Global gender disparities in science." *Nature*, *504*(7479), 211–213.

Latham, Gary P, Miriam Erez, and Edwin A Locke (1988). "Resolving scientific disputes by the joint design of crucial experiments by the antagonists: Application to the Erez–Latham dispute regarding participation in goal setting." *Journal of Applied Psychology*, *73*(4), 753–772.

Lazer, David and Allan Friedman (2007). "The network structure of exploration and exploitation." *Administrative Science Quarterly*, *52*(4), 667–694.

Lee, Carole J (2016). "eds Michael Brownstein and Jennifer Saul." *Implicit Bias and Philosophy, Volume 1: Metaphysics and Epistemology*, 265–282.

Leonard, Thomas C (2002). "Reflection on rules in science: An invisible-hand perspective." *Journal of Economic Methodology*, *9*(2), 141–168.

Leslie, Derek (2005). "Are delays in academic publishing necessary?" *American Economic Review*, *95*(1), 407–413.

Lewandowsky, Stephan and Klaus Oberauer (2020). "Low replicability can support robust and efficient science." *Nature Communications*, *11*(1), 1–12.

Lewandowsky, Stephan, Toby D Pilditch, Jens K Madsen, Naomi Oreskes, and James S Risbey (2019). "Influence and seepage: An evidence-resistant minority can affect public opinion and scientific belief formation." *Cognition*, *188*, 124–139.

Link, Albert N, Christopher A Swann, and Barry Bozeman (2008). "A time allocation study of university faculty." *Economics of Education Review*, *27*(4), 363–374.

Luukkonen, Terttu (2012). "Conservatism and risk-taking in peer review: Emerging ERC practices." *Research Evaluation*, *21*(1), 48–60.

Magnus, PD (2013). "What scientists know is not a function of what scientists know." *Philosophy of Science*, *80*(5), 840–849.

March, James G (1991). "Exploration and exploitation in organizational learning." *Organization Science*, *2*(1), 71–87.

Mason, Winter A, Andy Jones, and Robert L Goldstone (2008). "Propagation of innovations in networked groups." *Journal of Experimental Psychology: General*, *137*(3), 422–433.

Mayo, Deborah G (1996). *Error and the Growth of Experimental Knowledge*. University of Chicago Press.

Mayo, Deborah G (2018). "Statistical inference as severe testing." *Cambridge, UK: Cambridge Univ. Press Access provided by Katholieke Universiteit Leuven-KU Leuven on,*.

Mayo-Wilson, Conor, Kevin JS Zollman, and David Danks (2011). "The independence thesis: When individual and social epistemology diverge." *Philosophy of Science*, *78*(4), 653–677.

McDowell, John M and Janet Kiholm Smith (1992). "The effect of gender-sorting on propensity to coauthor: Implications for academic promotion." *Economic Inquiry*, *30*(1), 68–82.

McElreath, Richard and Paul E Smaldino (2015). "Replication, communication, and the population dynamics of scientific discovery." *Public Library of Science One*, *10*(8), e0136088.

McShane, Blakeley B, David Gal, Andrew Gelman, Christian Robert, and Jennifer L Tackett (2019). "Abandon statistical significance." *The American Statistician*, *73*(sup1), 235–245.

Merton, Robert K (1942). "A note on science and democracy." *Journal of Legal and Political Sociology*, *1*, 115–126.

Merton, Robert K (1957). "Priorities in scientific discovery: A chapter in the sociology of science." *American Sociological Review*, *22*(6), 635–659.

Merton, Robert K (1968). "The Matthew effect in science: The reward and communication systems of science are considered." *Science, 159*(3810), 56–63.

Merton, Robert K (1973). *The Sociology of Science: Theoretical and Empirical Investigations.* University of Chicago press.

Mohseni, Aydin (2023a). "HARKing: From misdiagnosis to misprescription." Working Paper.

Mohseni, Aydin (2023b). "Intervention and backfire in the replication crisis." Working Paper.

Mohseni, Aydin and Cole Randall Williams (2021). "Truth and conformity on networks." *Erkenntnis, 86,* 1509–1530.

Moonesinghe, Ramal, Muin J Khoury, and A Cecile JW Janssens (2007). "Most published research findings are false: But a little replication goes a long way." *Public Library of Science Medicine, 4*(2), e28. https://journals.plos.org/plosmedicine/article?id=10.1371/journal.pmed .0040028.

Muldoon, Ryan and Michael Weisberg (2011). "Robustness and idealization in models of cognitive labor." *Synthese, 183*(2), 161–174.

Munafò, Marcus R, Brian A Nosek, Dorothy VM Bishop, et al. (2017). "A manifesto for reproducible science." *Nature Human Behaviour, 1*(1), 1–9.

Murphy, Kevin R and Herman Aguinis (2019). "HARKing: How badly can cherry-picking and question trolling produce bias in published results?" *Journal of Business and Psychology, 34*(1), 1–17.

Newman, Mark EJ (2001). "Scientific collaboration networks: I. Network construction and fundamental results." *Physical Review E, 64*(1), 016131.

Newman, Mark EJ (2004). "Coauthorship networks and patterns of scientific collaboration." *Proceedings of the National Academy of Sciences, 101*(suppl 1), 5200–5205.

Nissen, Silas Boye, Tali Magidson, Kevin Gross, and Carl T Bergstrom (2016). "Publication bias and the canonization of false facts." *Elife, 5,* e21451.

Nosek, Brian A and Yoav Bar-Anan (2012). "Scientific utopia: I. Opening scientific communication." *Psychological Inquiry, 23*(3), 217–243.

Nosek, Brian A, Charles R Ebersole, Alexander C DeHaven, and David T Mellor (2018). "The preregistration revolution." *Proceedings of the National Academy of Sciences, 115*(11), 2600–2606.

Nosek, Brian A and Daniël Lakens (2014). "Registered reports: A method to increase the credibility of published results." *Social Psychology, 45*(3), 137–141.

Nosek, Brian A, Jeffrey R Spies, and Matt Motyl (2012). "Scientific utopia: II. Restructuring incentives and practices to promote truth over publishability." *Perspectives on Psychological Science, 7*(6), 615–631.

Nuzzo, Regina (2015). "Fooling ourselves." *Nature, 526*(7572), 182–185.

O'Connor, Cailin (2017). "The cultural red king effect." *The Journal of Mathematical Sociology, 41*(3), 155–171.

O'Connor, Cailin (2019a). "The natural selection of conservative science." *Studies in History and Philosophy of Science Part A, 76,* 24–29.

O'Connor, Cailin (2019b). *The Origins of Unfairness: Social Categories and Cultural Evolution.* Oxford University Press, USA.

O'Connor, Cailin and Justin Bruner (2019). "Dynamics and diversity in epistemic communities." *Erkenntnis, 84*(1), 101–119.

O'Connor, Cailin and James Owen Weatherall (2018). "Scientific polarization." *European Journal for Philosophy of Science, 8*(3), 855–875.

O'Connor, Cailin and James Owen Weatherall (2019). *The Misinformation Age.* Yale University Press.

Okasha, Samir (2006). *Evolution and the Levels of Selection.* Oxford University Press.

Olsson, E.J. (2013). A Bayesian Simulation Model of Group Deliberation and Polarization. In: Zenker, F. (eds.), *Bayesian Argumentation.* Synthese Library, vol 362. Springer: Dordrecht, 113–133. https://doi.org/10.1007/978-94-007-5357-0_6.

Open Science Collaboration, et al. (2015). "Estimating the reproducibility of psychological science." *Science, 349*(6251).

Oreskes, Naomi and Erik M Conway (2011). *Merchants of Doubt: How a Handful of Scientists Obscured the Truth on Issues from Tobacco Smoke to Global Warming.* Bloomsbury Publishing, USA.

Oster, Sharon (1980). "The optimal order for submitting manuscripts." *The American Economic Review, 70*(3), 444–448.

Peirce, Charles S (1967). "Note on the theory of the economy of research." *Operations Research, 15*(4), 643–648.

Pinto, Manuela Fernández and Daniel Fernández Pinto (2018). "Epistemic landscapes reloaded: An examination of agent-based models in social epistemology." *Historical Social Research/Historische Sozialforschung, 43*(1 (163)), 48–71.

Polanyi, Michael, John Ziman, and Steve Fuller (2000). "The republic of science: Its political and economic theory Minerva, I (1)(1962), 54–73." *Minerva, 38*(1), 1–32.

Popper, Karl R (1972). *Objective Knowledge.* Volume 360. Oxford University Press, Oxford.

Pöyhönen, Samuli (2017). "Value of cognitive diversity in science." *Synthese*, *194*(11), 4519–4540.

Protzko, John, Jon Krosnick, Leif D Nelson, et al. (2023). "High replicability of newly-discovered social-behavioral findings is achievable." Working Paper.

Radzvilas, Mantas, William Peden, and Francesco De Pretis (2021). "A battle in the statistics wars: A simulation-based comparison of Bayesian, Frequentist and Williamsonian methodologies." *Synthese*, *199*(5), 13689–13748.

Reijula, Samuli and Jaakko Kuorikoski (2019). "Modeling epistemic communities." In Fricker, Miranda, Peter J. Graham, David Henderson, and Nikolaj JLL Pedersen, (eds.), The Routledge Handbook of Social Epistemology. Routledge, 240–249.

Reijula, Samuli and Jaakko Kuorikoski (2021). "The diversity-ability trade-off in scientific problem solving." *Philosophy of Science*, *88*(5), 894–905.

Rogers, Everett M (2010). *Diffusion of Innovations*. Simon and Schuster.

Romero, Felipe (2016). "Can the behavioral sciences self-correct? A social epistemic study." *Studies in History and Philosophy of Science Part A*, *60*, 55–69.

Romero, Felipe (2017). "Novelty versus replicability: Virtues and vices in the reward system of science." *Philosophy of Science*, *84*(5), 1031–1043.

Romero, Felipe (2018). "Who should do replication labor?" *Advances in Methods and Practices in Psychological Science*, *1*(4), 516–537.

Romero, Felipe (2020). "The Division of Replication Labor." *Philosophy of Science*, *87*(5), 1014–1025.

Romero, Felipe and Jan Sprenger (2021). "Scientific self-correction: The Bayesian way." *Synthese*, *198*(23), 5803–5823.

Rosenstock, Sarita, Justin Bruner, and Cailin O'Connor (2017). "In epistemic networks, is less really more?" *Philosophy of Science*, *84*(2), 234–252.

Rosenthal, Robert (1979). "The file drawer problem and tolerance for null results." *Psychological Bulletin*, *86*(3), 638–641.

Rossiter, Margaret W (1993). "The Matthew Matilda effect in science." *Social Studies of Science*, *23*(2), 325–341.

Rubin, Hannah (2022). "Structural causes of citation gaps." *Philosophical Studies*, *179*, 2323–2345.

Rubin, Hannah and Cailin O'Connor (2018). "Discrimination and collaboration in science." *Philosophy of Science*, *85*(3), 380–402.

Rubin, Hannah and Mike D Schneider (2021). "Priority and privilege in scientific discovery." *Studies in History and Philosophy of Science Part A*, *89*, 202–211.

Rubin, Mark (2017). "When does HARKing hurt? Identifying when different types of undisclosed post hoc hypothesizing harm scientific progress." *Review of General Psychology, 21*(4), 308–320.

Rubin, Mark (2020). "Does preregistration improve the credibility of research findings?" *ArXiv Preprint ArXiv:2010.10513.*

Sampaio, Ricardo Barros, Marcus Vinicius de Araújo Fonseca, Fabio Zicker, et al. (2016). "Co-authorship network analysis in health research: Method and potential use." *Health Research Policy and Systems, 14*(1), 1–10.

Santana, Carlos (2021). "Let's not agree to disagree: The role of strategic disagreement in science." *Synthese, 198*(25), 6159–6177.

Sarsons, Heather (2017). "Recognition for group work: Gender differences in academia." *American Economic Review, 107*(5), 141–145.

Schneider, Mike D (2021). "Creativity in the social epistemology of science." *Philosophy of Science, 88*(5), 882–893.

Schneider, Mike D, Hannah Rubin, and Cailin O'Connor (2022). "Promoting diverse collaborations." In G. Ramsey and A. de Block (eds.), *The Dynamics of Science: Computational Frontiers in History and Philosophy of Science*. University of Pittsburgh Press.

Schofield, Paul N, Tania Bubela, Thomas Weaver, et al. (2009). "Post-publication sharing of data and tools." *Nature, 461*(7261), 171–173.

Selvin, Hanan C and Alan Stuart (1966). "Data-dredging procedures in survey analysis." *The American Statistician, 20*(3), 20–23.

Simmons, Joseph P, Leif D Nelson, and Uri Simonsohn (2016). "False-positive psychology: Undisclosed flexibility in data collection and analysis allows presenting anything as significant. *Psychological Science, 22*(11), 1359–1366

Singer, Daniel J (2019). "Diversity, not randomness, trumps ability." *Philosophy of Science, 86*(1), 178–191.

Smaldino, Paul (2019a). "Better methods can't make up for mediocre theory." *Nature, 575*(7783), 9–10.

Smaldino, Paul E (2019b). "Five models of science, illustrating how selection shapes methods." Working paper.

Smaldino, Paul and Cailin O'Connor (2022). "Interdisciplinarity can aid the spread of better methods between scientific communities." *Collective Intelligence, 1*(2), 26339137221131816.

Smaldino, Paul E (2019b). "Five models of science, illustrating how selection shapes methods." Working Paper.

Smaldino, Paul E and Richard McElreath (2016). "The natural selection of bad science." *Royal Society Open Science, 3*(9), 160384.

Smaldino, Paul E, Matthew A Turner, and Pablo A Contreras Kallens (2019). "Open science and modified funding lotteries can impede the natural selection of bad science." *Royal Society Open Science*, 6(7), 190194.

Smith, Adam (1759). *The Theory of Moral Sentiments*. Printed for A. Millar and A. Kincaid and J. Bell.

Smith, George Davey and Shah Ebrahim. "Data dredging, bias, or confounding: They can all get you into the BMJ and the Friday papers." *British Medical Journal 325*, 1437–1438.

Soderberg, Courtney K, Timothy M Errington, Sarah R Schiavone, et al. (2021). "Initial evidence of research quality of registered reports compared with the standard publishing model." *Nature Human Behaviour*, 5(8), 990–997.

Solomon, Miriam (2006). "Groupthink versus the wisdom of crowds: The social epistemology of deliberation and dissent." *The Southern Journal of Philosophy*, 44(S1), 28–42.

Sommers, Samuel R (2006). "On racial diversity and group decision making: Identifying multiple effects of racial composition on jury deliberations." *Journal of Personality and Social Psychology*, 90(4), 597–612.

Sonnert, Gerhard and Gerald Holton (1996). "Career patterns of women and men in the sciences." *American Scientist*, 84(1), 63–71.

Stanford, P Kyle (2019). "Unconceived alternatives and conservatism in science: The impact of professionalization, peer-review, and Big Science." *Synthese*, 196(10), 3915–3932.

Stephan, Paula (1996). *How Economics Shapes Science*. Harvard University Press.

Stewart, Alexander J and Joshua B Plotkin (2021). "The natural selection of good science." *Nature Human Behaviour*, 5, 1510–1518.

Strevens, Michael (2003). "The role of the priority rule in science." *The Journal of Philosophy*, 100(2), 55–79.

Strevens, Michael (2006). "The role of the Matthew effect in science." *Studies in History and Philosophy of Science Part A*, 37(2), 159–170.

Strevens, Michael (2011). "Economic approaches to understanding scientific norms." *Episteme*, 8(2), 184–200.

Strevens, Michael (2013). "Herding and the quest for credit." *Journal of Economic Methodology*, 20(1), 19–34.

Strevens, Michael (2017). "Scientific sharing: Communism and the social contract." In T. Boyer-Kassem, C. Mayo-Wilson, and M. Weisberg (eds.), *Scientific Collaboration and Collective Knowledge*, 3–33.

Szollosi, Aba, David Kellen, Danielle Navarro, et al. (2020). "Is preregistration worthwhile? *Trends in Cognitive Sciences*, 24(2), 94–95.

Tendeiro, Jorge N and Henk AL Kiers (2019). "A review of issues about null hypothesis Bayesian testing." *Psychological Methods*, *24*(6), 774–795.

Thagard, Paul (2006). "How to collaborate: Procedural knowledge in the cooperative development of science." *The Southern Journal of Philosophy*, *44*(S1), 177–196.

Thoma, Johanna (2015). "The epistemic division of labor revisited." *Philosophy of Science*, *82*(3), 454–472.

Thompson, Abigail (2014). "Does diversity trump ability?" *Notices of the AMS*, *61*(9), 1024–1030.

Tiokhin, Leo, Daniel Lakens, Paul E Smaldino, and Karthik Panchanathan (2021). "Shifting the level of selection in science." *Perspectives on Psychological Science*, 17456916231182568.

Tiokhin, Leonid and Maxime Derex (2019). "Competition for novelty reduces information sampling in a research game-a registered report." *Royal Society Open Science*, *6*(5), 180934.

Tiokhin, Leonid, Karthik Panchanathan, Daniel Lakens, et al. (2021). "Honest signaling in academic publishing." *Public Library of Science One*, *16*(2), e0246675.

Tiokhin, Leonid, Minhua Yan, and Thomas JH Morgan (2021). "Competition for priority harms the reliability of science, but reforms can help." *Nature Human Behaviour*, *5*(7), 857–867.

Viola, Marco (2015). "Some remarks on the division of cognitive labor." *RT. A Journal on Research Policy and Evaluation*, *3*.

Wagner, Elliott and Jonathan Herington (2021). "Agent-based models of dual-use research restrictions." *The British Journal for the Philosophy of Science*, *72*(2), 377–399.

Watts, Christopher and Nigel Gilbert (2011). "Does cumulative advantage affect collective learning in science? An agent-based simulation." *Scientometrics*, *89*(1), 437–463.

Weatherall, James Owen and Cailin O'Connor (2021a). "Conformity in scientific networks." *Synthese*, *198*(8), 7257–7278.

Weatherall, James Owen and Cailin O'Connor (2021b). "Endogenous epistemic factionalization." *Synthese*, *198*(25), 6179–6200.

Weatherall, James Owen, Cailin O'Connor, and Justin P Bruner (2020). "How to beat science and influence people: Policymakers and propaganda in epistemic networks." *The British Journal for the Philosophy of Science*, *71*(4), 1157–1186.

Weisberg, Michael and Ryan Muldoon (2009). "Epistemic landscapes and the division of cognitive labor." *Philosophy of Science*, *76*(2), 225–252.

West, Jevin D, Jennifer Jacquet, Molly M King, Shelley J Correll, and Carl T Bergstrom (2013). "The role of gender in scholarly authorship." *Public Library of Science One*, *8*(7), e66212.

Wright, Sewall (1932). "The roles of mutation, inbreeding, crossbreeding, and selection in evolution." Proceedings of the Sixth International Congress on Genetics, 356–366.

Wu, Jingyi (forthcoming). Better than Best. *Philosophy of Science*.

Wu, Jingyi (2023a). "The communist norm and industrial science." Withholding Knowledge. Working Paper.

Wu, Jingyi (2023b). "Epistemic advantage on the margin: A network standpoint epistemology." *Philosophy and Phenomenological Research*, *106*(3), 755–777.

Wu, Jingyi and Cailin O'Connor (2023). "How should we promote transient diversity in science? *Synthese*, *201*(2), 37.

Wu, Jingyi, Cailin O'Connor, and Paul E Smaldino (forthcoming). "The cultural evolution of science." In J. Tehrani, R. Kendal and J. Kendal (eds.), *The Oxford Handbook of Cultural Evolution*. Oxford University Press.

Xie, Yu, Kai Wang, and Yan Kong (2021). "Prevalence of research misconduct and questionable research practices: A systematic review and meta-analysis." *Science and Engineering Ethics*, *27*(4), 1–28.

Yahosseini, Kyanoush Seyed and Mehdi Moussaïd (2020). "Comparing groups of independent solvers and transmission chains as methods for collective problem-solving." *Scientific Reports*, *10*(1), 1–9.

Zollman, Kevin (2023). "The scientific ponzi scheme." Working Paper.

Zollman, Kevin JS (2007). "The communication structure of epistemic communities." *Philosophy of Science*, *74*(5), 574–587.

Zollman, Kevin JS (2009). "Optimal publishing strategies." *Episteme*, *6*(2), 185–199.

Zollman, Kevin JS (2010). "The epistemic benefit of transient diversity." *Erkenntnis*, *72*(1), 17–35.

Zollman, Kevin JS (2018). "The credit economy and the economic rationality of science." *The Journal of Philosophy*, *115*(1), 5–33.

Acknowledgements

I would like to thank my long-time collaborator and partner James Owen Weatherall for both personal and professional support in this project. Many thanks to our NSF working group for feedback and discussion including Clara Bradley, Matthew Coates, David Freeborn, Nathan Gabriel, Ben Genta, Daniel Herrmann, and Jingyi Wu. Special thanks to folks who gave comments or feedback on drafts of this Element: Ben Genta, Liam Kofi Bright, Remco Heesen, Aydin Mohseni, Hannah Rubin, Paul Smaldino, and Jingyi Wu. Thanks to two anonymous reviewers, one of whom was obviously Kevin Zollman, for extensive comments and feedback.

Cambridge Elements ☰

Philosophy of Science

Jacob Stegenga
University of Cambridge

Jacob Stegenga is a Reader in the Department of History and Philosophy of Science at the University of Cambridge. He has published widely on fundamental topics in reasoning and rationality and philosophical problems in medicine and biology. Prior to joining Cambridge he taught in the United States and Canada, and he received his PhD from the University of California San Diego.

About the Series

This series of Elements in Philosophy of Science provides an extensive overview of the themes, topics and debates which constitute the philosophy of science. Distinguished specialists provide an up-to-date summary of the results of current research on their topics, as well as offering their own take on those topics and drawing original conclusions.

Cambridge Elements ≡

Philosophy of Science

Printed in the United States
by Baker & Taylor Publisher Services

Printed in the United States
by Baker & Taylor Publisher Services